D0938641

POLYMERS
AS BIOMATERIALS

POLYMERS
AS BIOMATERIALS

Edited by

Shalaby W. Shalaby

ETHICON, Inc.
A Johnson & Johnson Company
Somerville, New Jersey

Allan S. Hoffman
Buddy D. Ratner
and
Thomas A. Horbett

University of Washington
Seattle, Washington

PLENUM PRESS • NEW YORK AND LONDON

Library of Congress Cataloging in Publication Data

Main entry under title:

Polymers as biomaterials.

"Proceedings of a symposium on polymers as biomaterials, held March 22–25, 1983, at the American Chemical Society meeting in Seattle, Washington"—T.p. verso.
   Bibliography: p.
   Includes index.
   1. Polymers in medicine—Congresses. 2. Polymers in medicine—Physiological effect—Congresses. 3. Colloids in medicine—Congresses. 4. Drugs—Vehicles—Congresses. I. Shalaby, Shalaby W. II. American Chemical Society. Meeting (185th: 1983: Seattle, Wash.)
R857.P6P6 1985   610′.28   84-20666
ISBN 0-306-41886-X

Proceedings of a symposium on Polymers as Biomaterials, held March 22–25, 1983, at the American Chemical Society meeting in Seattle, Washington

© 1984 Plenum Press, New York
A Division of Plenum Publishing Corporation
233 Spring Street, New York, N.Y. 10013

Printed in the United States of America

PREFACE

Nearly 4000 years ago, the Egyptians used linen, a natural polymeric material, for suturing wounds. About 600 B.C., the Indians used other forms of natural polymers such as cotton, horse hair, and leather in repairing wounds. Wound closure procedures using silk sutures, based mostly on polypeptides, are likely to have been practiced during the second century.

Surgical application of natural polymers continued to represent the major use of polymers until the twentieth century. Not too long after the development of several major synthetic polymers, their use in biomedical applications has attracted the attention of many researchers and clinicians. Over the past few years, interest in the biomedical applications of polymers has grown considerably. This has been the result of the inevitable collaborative efforts of innovative materials scientists, engineers and clinicians. The establishment of the Society for Biomaterials, in our opinion, catalyzed the growing interest in the use of polymers for biomedical application.

In a major effort to bring team players even closer, a five-day symposium on "Polymers as Biomaterials" was held in Seattle, Washington, in March, 1983 as part of the national meeting of the American Chemical Society. The symposium was designed to provide a forum for communicating technical and clinical data to colleagues with a broad spectrum of interest in the biomedical applications of polymers.

This volume represents most of the symposium proceedings and consists of six sections. Section A describes new and known polymeric materials and their physico-chemical properties as they pertain to specific biomedical applications. In Section B, some experimental aspects of surface characterization and interaction of blood components with polymeric surfaces are documented. Interaction of polymeric materials with the biologic environment is used in Section C to demonstrate the response of polymeric materials to changes in the surrounding environment and biodegradation of synthetic polymers. Section D encompasses key interactions of the biologic environment with polymeric surfaces. Thus, the effects of surface properties on

v

the behavior of living cells and certain cellular events are em-
phasized. Section E is dedicated to the role of polymers in the
development of new, controlled drug delivery systems. The final
section of this volume addresses the release kinetics and transport
through hydrogels and the interaction of their hydrophilic surfaces
with blood components and ionic species of the biologic environment.

This book is designed and certainly hoped to be a valuble
asset to investigators in academic, governmental, and industrial
institutions who intend to be at the forefront of new technologies
relevant to the development of new polymeric materials for known
and novel biomedical applications.

S.W. Shalaby
A.S. Hoffman
B.D. Ratner
T.A. Horbett

CONTENTS

SECTION A
MATERIALS & PROPERTIES

SECTION E
DRUG DELIVERY SYSTEMS

SECTION F
HYDROGELS

POLY(β-MALIC ACID) AS A SOURCE OF POLYVALENT DRUG CARRIERS:

POSSIBLE EFFECTS OF HYDROPHOBIC SUBSTITUENTS IN AQUEOUS MEDIA

Christian Braud and Michel Vert

Université de Rouen, ERA CNRS n° 471
LSM, INSCIR, BP 8, 76130 Mont Saint Aignan, France

INTRODUCTION

According to literature, the interest in investigating synthetic macromolecular compounds for uses in biology and medicine is
rapidly increasing. On one hand, attention has been focussed on
polymer based implants for prostheses, tissue restauration and
drug-delivery devices. In these cases, macromolecular compounds act,
and sometimes react, as solids in the biological milieu. However,
attention has been also payed to uses of synthetic polymers molecu-
larly dispersed in body fluids and, more generally, in the living
medium. Two types of compounds have been considered : macromolecules
that are pharmacologically active by themselves (polymeric drugs),
and drug-polymeric carrier systems which bear drugs temporarily
attached to the polymer backbone either covalently (macromolecular
prodrugs) or because of physical interactions. It is of value to
note that distinction between polymeric drugs and macromolecular
prodrugs is based on expected uses. Indeed, it is likely that no
macromolecule is inert in the living medium.

A great number of biologically active macromolecular systems
have been considered in the last three decades (1-7). However, the
literature reveals primarily chemistry work, for the synthesis of
of their properties. It is remarkable that only little has been made
to investigate structure-activity relationships in detail though
many authors early suspected the importance of structural features
such as chain coiling and/or folding, tacticity, electrolytic
character etc... (Donaruma (1), Kopecek (4), Ottenbrite (7)) in drug
design work. In this respect, not very much more than what was
known in 1974 (1) is available at the moment (8). There are many

1

reasons for the lack of the physico-chemical characterizations needed to acheive significant structural identification of biologically active synthetic polymers. One can mentioned : -limited amounts of available samples, -difficulties to obtain comparable data from various techniques that require different experimental conditions, -complexity of the macromolecular systems and of the biological milieu, and also -the race to find out new functional species through fast-screening.

As already pointed out by Ringsdorf (2) and others (6,7), biologically active polymers have to be tailored to fit the desired biological and physical properties. Water-solubility is one of the prerequisites of this type of polymers if they are to be injected in body fluids (2,5). A mean to water-solubilize macromolecules consists in the introduction of hydrophilic groups (4) within polymeric chain or side-chains, in particular ionic groups (9,10). Many of the systems reported in literature are of the polyanion or polycation types, though authors seldom comment on this particularity and on the special conformational properties related to it. More attention has been payed to other prerequisites such as non-toxicity of the prodrug, non-immunogenicity for repetitive administrations, presence of cleavable bonds and possibly spacer arms, degradability and resorbability of the carrier-backbone in order to prevent accumulation in the body, etc... . In spite of these efforts, one must admit that there still is a number of unsolved problems in the field. The use of synthetic soluble drug carriers in human medicine is regarded sceptically by the medical profession, partly because of the risk of deposition and accumulation of high molecular weight compounds in the body, and by the pharmaceutical profession, because each system has to be regarded as a new drug and, as such, can be the source of considerable expense before the first effective use.

In Rouen, we became interested in water-soluble polymers of potential pharmacological importance because of our experience in three different domains of the polymer science i.e. -synthesis of sophisticated optically active polymers (11), -study of the conformational behavior of hydrophilic-hydrophobic polyelectrolytes (12-14) and -use of bioresorbable polymers for biomedical applications (15).

In a joined program with the University of Massachusetts (16), we selected a new polyacid, poly(β-malic acid)(PMLA 100), as a suitable candidate not only to be a sucessful polyvalent drug-carrier i.e.fulfilling the prerequisites as much as possible, but also to be a compound of choice for the study of structure - physico-chemical properties - biological activity relationships in regard to injection in body fluids. Poly(β-malic acid) was still unknown, but a route was found to synthesize the racemic compound (16). The necessary efforts were done for various reasons. Firstly, chains of PMLA 100 include L-malic acid repeating units, a well-known metabolite . Secondly, carboxylic acid groups are present as pendant groups that

2

could water-solubilize the polymeric chains at neutral and provide means to attach drugs and other required substituents. Thirdly, PMLA 100 macromolecules have labile ester bonds in the main chain which might thus be biodegradable in the body (17).

$$(-O-CH-CH_2-CO-)_n$$
$$|$$
$$COOH$$

PMLA 100

According to the preceeding remarks, the goal of this paper is threefold. Firstly, we want to comment on the modifications of physico-chemical properties and on the possible resulting perturbations of biological properties of macromolecular biologically active polymers when the proportion and the position of attached substituents are changed either at the synthesis stage or during the release of the active species. For this discussion, physico-chemical data are considered of two series of homologous polyacids with variable proportions of hydrophilic and hydrophobic substituents - respectively poly(acrylic acid co-N-N-disubstituted acylamide) and poly(maleic acid-alt-alkylvinyl ether). In a second part, we will present some data obtained for partially benzylated PMLA 100 that support the general comments. At last, we will show that some benzylated PMLA 100 derivatives for lipophilic microdomains in water which might be usable as microemulsion-like systems for trapping hydrophobic drugs through physical interactions (18).

GENERAL COMMENT ON POSSIBLE SOLUTION BEHAVIORS OF CHARGED BIOLOGICALLY ACTIVE POLYMERS IN AQUEOUS MEDIA

Most of the soluble polymeric systems that have been tailor-made for pharmacological purposes bear hydrophobic substituents to a certain extent. From our experience based on various hydrophilic-hydrophobic polyelectrolytes of the carboxylic type, we know that a small change in the balance of hydrophilic and hydrophobic segments along a hydrophilic chain can affect the conformational structure of the whole macromolecule (fig.1). For a hydrophilic polyelectrolyte, an increase in hydrophobicity can generate a folding of the macromolecules leading to more or less compact conformations (hypercoiled or polysoap-like) at given ionic strength and degree of ionization of -COOH groups imposed by pH. Excessive hydrophobicity can result in aggregation and, at the limit, in non-solubility.

The effect of the amount of hydrophobic moieties, randomly attached to hydrophilic polycarboxylic acid chains, on conformations can be exemplified by considering a related series of random copolymers (COP Y) with acrylic acid moieties (hydrophilic) and N-(sec-butyl-N-methyl acrylamide moieties (hydrophobic) whose chemical composition is defined by the percentage (Y) of acidic units along the chain (19).

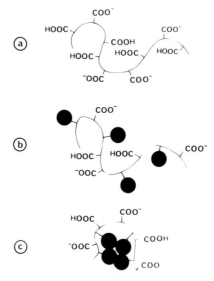

Fig. 1 Schematic representation of the effects of the substitution of hydrophilic groups (-COOH and -COO⁻) by hydrophobic ones on the conformational structures of a polyelectrolytic macromolecule. In (a), the well-solvated chain takes on an open-coil conformation. The substitution of some hydrophilic groups by hydrophobic ones (black dots) can preserve the open-coil structures (b) or can induce a compact-coil structure (c) depending on the overall composition.

In these copolymers, the N-(sec-butyl)-N-methyl acrylamide moieties can be regarded as a model of a neutral hydrophobic drug attached to carboxylic groups of poly(acrylic acid) through amide bond.

$$(-CH-CH_2)_m ----\mid---- (-CH-CH_2)_p$$

$$\begin{array}{ccc} \mid & & \mid \\ COOH & & CO \\ & & \mid \\ & & H_3C-N-CH-CH_3 \\ & & \mid \\ & & C_2H_5 \end{array}$$

$$\underline{COP\ Y}\ with\ Y\ =\ 100.\frac{m}{m+p}$$

A few years ago, the conformational behavior of COP Y copolymers in salt-free water was investigated at different pH (12). Some of the data are schematically recalled in Figure 2. At low pH values corresponding to self-ionization, copolymers with low Y values are water-insoluble, those with high Y values are typically hydrophilic, while the intermediate ones take on compactly coiled conformations. Basically, the rise of the pH of salt-free COP Y solutions from the

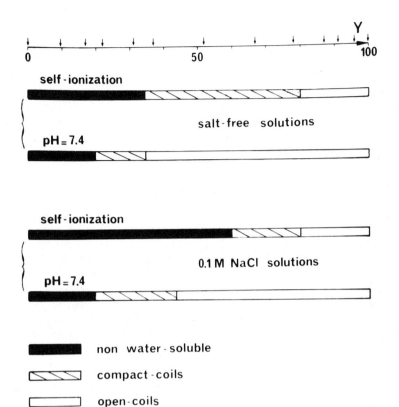

Fig. 2   Behaviors of COP Y in aqueous solutions as a function of
composition (Y). Solutions at self-ionization ($\simeq 1$ g/l) were
obtained by percolating solutions of COP Y in Na$^+$ form through
a cation-exchange resin column under the H$^+$ form (19). Solutions at pH = 7.4 were obtained by neutralizing the self-
ionized solutions by NaOH. Conformational structures in
salt-free solutions were deduced from potentiometric titration
curves (12,13). Conformational structures in 0.1 M NaCl solutions were deduced from the ability of COP Y solutions to
water-solubilize Yellow OB, a dye insoluble in water (14).
Arrows on the Y axis point the composition of the different
copolymers studied.

values due to self-ionization to 7.4, the physiological pH, enhances
the hydrophilicity of the whole systems as charged −COO$^-$ groups are
more hydrophilic than COOH ones. Figure 2 shows that composition
ranges corresponding to water-insolubility and to compact-coiling
are narrowed while the open-coil range is enlarged.

On the other hand, the presence of salt (0.1 M NaCl) slightly
modifies these features. Indeed, the addition of salt to aqueous

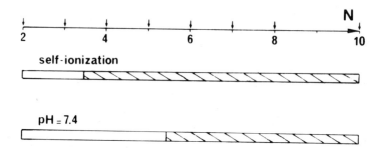

Fig. 3   Conformational structures of salt-free solutions of MAc/RVE
         copolymers as a function of the number of carbon atoms (N)
         of the R substituent. Meanings of the symbols are the same
         as in Figure 2.

solutions of a polycarboxylic acid increases the self-ionization of
-COOH groups and screens the -COO⁻ charges. Ionized COOH groups tend
to increase the hydrophilicity of the whole system while the screening of electric charges acts oppositely. Data in Figure 2 show that
screening effects dominate as the water-insolubility and the compact
coil ranges enlarge at the expense of the open-coil one for both
self-ionization and natural pH values.

     From the behavior of members of the COP Y series, one can
already conclude that attachment of hydrophobic drugs to a polycarboxylic acid type carrier may cause small effects or give rise to
dramatic conformational changes or even water-insolubility depending
on the hydrophilic-hydrophobic balance of the whole system.

     COP Y copolymers have allowed us to show that the amount of
hydrophobic substituents present in macromolecular chains should be
carefully considered when one wants to tailor-make a complex system
such as a biologically active polymer. We are now going to show that
the location of substituents along a macromolecular chain can also
be of importance for the solution behavior of polyacid   macromolecules. Both the effect of hydrophobic substituents and the effect of
their position can be exemplified by comparing the case of COP Y
random copolymers with the case of alternating copolymers of maleic
acid and alkylvinylethers (MAc/RVE) with various alkyl groups
(R = ethyl, n-propyl, n-butyl, iso-butyl, tert-butyl, 1-methylpropyl,
1-methylbutyl, 2-methylbutyl, 3-methylpentyl, 4-methylhexyl and
1-methylheptyl).

     Figure 3 shows the schematic representation of the solution
behavior of these polyacids in regard to the number of carbon atoms
(N) of the alkyl moieties attached to the polymeric chains through
ether side-bonds. Minor effects due to chain isomerism in these
alkyl groups have been found (14), but are not taken into account

here. In this MAc/RVE series, all the members are water-soluble in salt-free water whatever the pH and the size of the alkyl groups. For the COP Y random copolymers, we have seen before that there are composition ranges where the presence of hydrophobic substituents causes water-insolubility. Similar behaviors have been observed for many other random hydrophilic-hydrophobic polycarboxylic acids (20-22). The absence of water-insolubility for highly hydrophobic alternating MAc/RVE copolymers (N>8) suggests that blocks of several hydrophobic repeating units are necessary to cause macroscopic water-insolubility.

Anyhow, at the molecular scale, the size of alkyl substituents does affect the conformational behavior in some extent since increasing N causes conformations to change from open-coils to compact-coils due to increasing hydrophobicity (14,23). The critical alkyl side-chain length was found in the C_3-C_4 range under self-ionization conditions and was found in the C_5-C_6 range at pH = 7.4 (14). This shift is primarily caused by the ionization of acidic groups.

The combination of the features shown by the two series of compounds selected as examples for this discussion demonstrates that the tailor-making of sophisticated polyelectrolytic chains can result in totally different conformational structures in aqueous solutions depending not only on the nature and the amount of the various substituents but also on their position in regard to each other and to the polymeric backbone. Accordingly, the solution behavior related to the presence of hydrophobic and hydrophilic groups in therapeutic polymers can result in unexpected property changes (biocompatibility, bioavailability) due to sudden composition-induced modifications in conformations and, thus, in the accessibility of the system to active sites and in interactions with the components of the living medium. Because it is still impossible to quantify the hydrophilic-hydrophobic balance of a polymeric system, the chemical modification of a polyacid chain can be tricky as similar behaviors can be obtained in large ranges of composition while, in other ranges, sharp conformational changes can be observed for minor chemical modifications.

The literature did not contain any examples showing identified biological effects due to conformational structure changes. An example has been found in the series of PMLA 100 derivatives.

EFFECT OF HYDROPHOBIC SUBSTITUENTS AND OF THEIR DISTRIBUTION ON ACUTE TOXICITY OF PMLA 100 DERIVATIVES

PMLA 100 is a typical hydrophilic polyelectrolyte which is characterized by a moderately weak carboxylic acidity (pK_a at half-neutralization = 4.4 for a solution of $\sim 5.10^{-2}$ eq/l (24)). It must be noted that at neutral all the carboxylic groups are ionized.

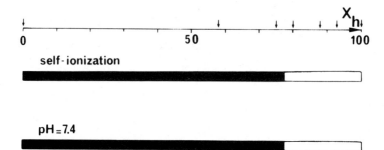

Fig. 4 Solubility in aqueous medium of salt-free solutions of PMLA X_h as a function of composition (X_h). Black aera means water-insolubility while white aera means macroscopic solubility.

$$(-O-CH-CH_2-CO-)_n$$
$$COOBz$$

PMLABE

$$H_2 \downarrow Pd/C$$

$$(-O-CH-CH_2-CO-)_m ----\mid---- (-O-CH-CH_2-CO-)_p$$
$$COOBz COOH$$

PMLA X_h

$$(X_h = 100 \cdot \frac{p}{m+p})$$

$$H_2 \downarrow Pd/C$$

$$(-O-CH-CH_2-CO-)_n$$
$$COOH$$

PMLA 100

The first route to PMLA 100 was reported a few years ago (16,25). It is based on the catalytic hydrogenolysis (Pd charcoal) of the parent benzyl ester (PMLABE) and goes through a series of acid-benzyl ester copolymers (PMLA X_h) with increasing X_h values (X = percentage of acidic repeating units and subscript h for copolymers obtained by catalytic hydrogenolysis).

Since side-chains of the benzyl ester-type are known to behave hydrophobically in the case of poly(methacrylic acid) (21), we initially tried to take advantage of the benzyl groups present in intermediate PMLA X_h copolymers to investigate composition - physico-chemical - biological property relationships in a simple case. Our investigations were facilitated later because PMLA X_h copolymers rapidly revealed unusual structural features which allowed us to extend our goal to the exemplification of repeating unit distribution - physico-chemical property - acute toxicity relationships.

Based on our experience recalled previously, composition-dependence was expected for the polyelectrolytic characteristics of PMLA X_h copolymers. We effectively found water-insolubility for $X_h < 75$. However, the following features were also observed (24) :

8

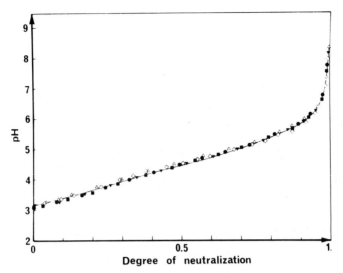

Fig. 5   Potentiometric titration curves of PMLA 100 (●), PMLA 93_h (■),
PMLA 88_h (▼), PMLA 75_h (△) and PMLA 58_h (◇) in a 13-87
V/V acetone-water mixture ($3.8 \ 10^{-2}$N < C_{acid} < $5.3 \ 10^{-2}$N).

　　　　i) the same critical composition was found for both the
salt form (pH = 7.4) and the acid form (Fig.4). Thus, the classical
percolation technique (20) could not be used to condition the acid
form as usual (12,14). In the presence of excess NaOH, PMLA 75_h and
PMLA 58_h behaved as an ion-exchange resin of the carboxylic type
instead of going into solution.
　　　　ii) the acidic form of the PMLA X_h copolymers conditioned
by addition of water to acetone solutions up to the same 13-87 V/V
acetone-water composition showed a single potentiometric titration
curve (Fig. 5) whatever the macroscopic aspect of the mixtures (milky
for X_h = 58, turbid for X_h = 75 and water-clear for X_h > 75). This
feature is unusual when compared to what is observed for hydrophilic-
hydrophobic polyelectrolytes (13).

　　　These findings suggested that PMLA X_h copolymers have a non-random copolymer structure i.e. that acid and benzyl ester repeating
units are not randomly distributed, as in the case of COP Y copolymers, for example. Actually, we have recently assumed that PMLA X_h
copolymers have a A-B block structure probably formed during the
hydrogenolysis reaction (24). Indeed, the water-soluble members of
the PMLA X_h series exhibit typical characteristics of micelle-forming
amphiphilic bi-block copolymers (26). For example, PMLA 80_h ($\overline{M}_{GPC}$ ≃
15,000 in organic solvents) shows very high molecular weight in
aqueous media ($\overline{M}_{GPC}$ > 200,000). Furthermore, PMLA 80_h forms hydrophobic micro-domains in aqueous media that can dissolved Yellow OB even
at neutral (27). Though the block structure of PMLA 80_h is not yet
totally elucidated, it offered us the chance to evaluate the effect

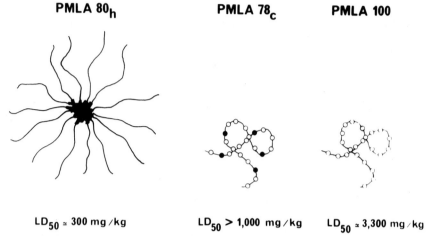

PMLA 80$_h$ PMLA 78$_c$ PMLA 100

LD$_{50}$ ≈ 300 mg/kg   LD$_{50}$ > 1,000 mg/kg   LD$_{50}$ ≈ 3,300 mg/kg

Fig. 6   Structure – acute toxicity relationship for the sodium salts of PMLA 80$_h$, PMLA 78$_c$ and PMLA 100. ～ and ● figure the benzyl ester block and the individual benzyl ester moiety respectively, and ～ and ○ figure the acid block and the individual acid moiety.

of repeating unit distribution on biological properties in a given series of compounds. A random copolymer PMLA 78$_c$, where c stands for coupling, was prepared by chemical coupling of benzyl alcohol to PMLA 100 using DCC.

$$(-O-CH-CH_2-CO-)_n \qquad\qquad \text{PMLA 100}$$
$$\qquad\quad | $$
$$\qquad\quad COOH$$

$$DCC \downarrow \text{benzyl alcohol}$$

$$(-O-CH-CH_2-CO)_m ----|----(-O-CH-CH_2-CO-)_p \qquad \text{PMLA X}_c$$
$$\qquad | \qquad\qquad\qquad\qquad\qquad\qquad | $$
$$\qquad COOBz \qquad\qquad\qquad\qquad\qquad COOH$$
$$(X_c = 100 \cdot \frac{p}{m+p})$$

As expected, this copolymer did not show any evidence for micelle formation, in agreement with random distribution of repeating units (27).

Acute toxicity, which is the simplest test of relevant interest for practical approaches have, thus, been considered for both copolymers and for PMLA 100 using mice as testing animals. Compounds were given i.p. to lots of Mus Musculus mice weighing 20 to 25 g each.

Macromolecular structures and toxicity data are schematically represented in Figure 6. The following features can be deduced from these experiments :

- toxicity of open-coiled PMLA 100 is very low ($LD_{50} \simeq 3.3$ g/kg of body weight),
- toxicity of open-coiled PMLA 78_c remains low ($LD_{50} > 1$ g/kg),
- toxicity of aggregated PMLA 80_h increases very much as LD_{50} drops to 300 mg/kg.

These data show that PMLA 100 is almost non-toxic. The comparison between values of the LD_{50} found for PMLA 78_c and PMLA 80_h suggests that the non-random distribution of PMLA 80_h repeating units is responsible for the difference in toxicity since both copolymers have the same gross composition in hydrophilic acid and hydrophobic benzyl ester substituents. It is likely that the excess toxicity shown by PMLA 80_h in regard to PMLA 78_c is due to the presence of micelles. It is of value to point out that even if PMLA 80_h appears more toxic than PMLA 100 and PMLA 78_c, its LD_{50} is still high in regard to many toxic drugs. Data obtained so far for PMLA 78_c suggest that the toxicity does not decrease very much in regard to the non-substituted parent polyacid when the conformational structure remains of the open-coil type.

At this point of the discussion, PMLA 100 appears as a worthwhile potential carrier for many drugs that may be attached through carboxylic groups. This statement is further supported by the fact that no immune response has been identified, so far, even after repeated i.p. injections in rabbits. However, the interest of the polyvalency of such a carrier may be affected by the covalent attachment of pharmacological compounds. Indeed, each combination can constitute a new biologically active compound whose properties have to be evaluated if effective application in human is expected. It is probably one of the major reason for pharmaceutical people to prefer drug delivery systems based on embedding or microencapsulation within biodegradable or bioerodible polymer matrices to acheive time-controlled releasing of drugs (28). In this regard, the micellar structure of PMLA X_h appears of special interest for water-solubilization of lipophilic drugs in aqueous media, a present-day problem approached through microemulsions and phopholipidic vesicules (29). Investigations have been undertaken to evaluate the exact potential of PMLA X_h micelles for temporary drug-trapping.

SOLUBILIZATION OF LIPOPHILIC DRUGS IN PMLA 80_h AQUEOUS DISPERSIONS

Hydrophobic microdomain-forming polyelectrolytes are well known for their ability to favor the dissolution of aromatic hydrocarbons (30) or water-insoluble dyes (14,23) in aqueous media. The plurimolecular micelles formed by PMLA 80_h constitute a special case of hydrophobic microdomains in water solutions that can dissolve the Yellow OB dye (27). Such micelles can also immobilize poorly water-soluble therapeutic agents. An example is given here using progesterone. This compound is very sparingly soluble in water

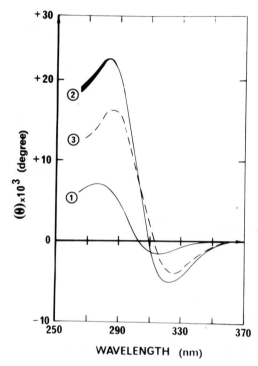

Fig. 7 Circular dichroism spectra of saturated solutions of progesterone in water (curve 1) and in water with 10 mg / ml PMLA 80$_h$ (curve 2). Curve 3 results from the difference between curve 2 and curve 1. It shows the CD contribution of progesterone dissolved in the hydrophobic core of the PMLA 80$_h$ macromolecular micelles.

($\simeq$ 12 µg/ml at 25 °C) and is usually injected under the form of oil-in-water microemulsions.

The solubility of progesterone in water and in aqueous solutions of PMLA 80$_h$ was monitored by circular dichroism in the range of 350-250 nm where this optically active molecule shows Cotton effects. Typical data are presented in Figure 7. Without polymer (curve 1), the CD spectrum of progesterone is characterized by a bisignate curve that denotes the presence of two Cotton effects of opposite sign, at least. The bisignate curve crosses over the base-line at λ = 302.5 nm. The concentration of dissolved progesterone was deduced from the comparison of CD signals obtained for aliquots diluted with methanol so that the final solvent composition be 80-20 V/V (methanol-water) with signals of standards containing known amounts of progesterone in the same solvent system. The concentration of progesterone was found much larger in saturated water containing PMLA 80$_h$ at the LD$_{50}$ concentration (10 mg / ml) than in saturated pure water, namely 50 µg / ml .

In order to determine the CD characteristics of the progesterone
dissolved in the micelles, the CD contribution due to progesterone
normally dissolved in the water-phase was substracted from curve $\underline{2}$
assuming that the presence of the macromolecules does not affect the
solubility of progesterone in water. Accordingly, the resultant curve
(curve 3) corresponds to progesterone solubilized in excess (38
$\mu g/ml$) and localized in the hydrophobic core of the micelles.
be noted that the cross-over of the base-line was red-shifted to
313 nm similarly to what is observed when moving progesterone from
water to organic solvents (ethanol, methanol or dioxane). This
feature bears out the presence of the organic core of PMLA 80_h
micelles. Obviously, a much larger solubilizing effect can be obtained
for concentrations higher than that corresponding to the LD_{50}.

Further experimentation will show whether such micellar systems
are stable enough to retain temporarily hydrophobic compounds and
allow their slow release after injection in the body.

CONCLUSIONS

In this paper, we have recalled that poly(β-malic acid), a
hydrophilic polycarboxylic acid, appears more and more as a polyvalent
drug carrier. It can be certainly used, as expected initially, to
tailor-make biologically active polymers by the attachment of selected
substituents to some of the carboxylic groups. However, we have shown
that the tailoring has to be made carefully, especially if hydrophobic substituents are involved. Indeed the amount and the distribution
of these substituents can be important factors through their possible
effects on the conformational structures, and consequently on the
biological properties of the resulting systems. On another hand, we
have also shown that the particular micellar structures taken on by
poly(β-malic acid) benzyl ester derivatives obtained by catalytic
hydrogenolysis might be of special interest to temporarily immobilize
lipophilic drugs, and thus to act as a polyvalent drug carrier of
the microemulsion- or vesicule-types.

However, we keep in mind that many compounds which are present
in the living medium can also interact with synthetic polymers either
through adsorption or through complexation causing further perturbations of the conformational structures of these drug - polymeric
carrier systems.

ACKNOWLEDGEMENTS

The authors are indebted to P. Bouffard and Dr M. Clabaut of the
department of Biology of the University of Rouen for performing
toxicity tests on mice, and to Dr B. Delpech of the Becquerel Cancer
Institute of Rouen for immunologic assays.

REFERENCES

1. L. G. Donaruma, Synthetic biologically active polymers, Progr. Polymer Sci. 4:1 (1974)
2. H. Ringsdorf, Pharmacologically active polymers, J. Polym. Sci., Polym. Symp. 51:135 (1975)
3. H. G. Batz, Polymeric drugs, Adv. Polym. Sci. 23:25 (1977)
4. J. Kopecek, Soluble biomedical polymers, Polym. in Med. 7:191 (1977)
5. C. G. Gebelein, Survey of chemotherapeutic polymers, Polym. News 4:163 (1978)
6. C. M. Samour, Polymeric drugs, Chemteck 8:494 (1978)
7. R. M. Ottenbrite, Introduction to polymers in biology and medicine, in: "Anionic Polymeric Drugs," L. G. Donaruma, R. M. Ottenbrite and O. Vogl, eds., J. Wiley, New York (1980)
8. M. Vert, Ionizable polymers for temporary uses in medicine, in: "Proceedings of the 28th IUPAC Symposium on Macromolecules," Amherst (1982)
9. R. M. Ottenbrite, Structure and biological activities of some polyanionic polymers, in: "Anionic Polymeric Drugs," L. G. Donaruma, R. M. Ottenbrite and O. Vogl, eds., J. Wiley, New York (1980)
10. P. A. Kramer, Synthetic polyelectrolytes as modifiers of drug disposition and effectiveness : an overview, in: "Optimization of Drug Delivery," H. Bundgaard, A. Bagger Hansen and H. Kofod, eds., Munksgaard, Copenhagen (1982)
11. M. Vert, Optical activity of reactive "non-regular" synthetic polymers : properties and applications, in: "Charged and Reactive Polymers - Optically Active Polymers," E. Sélégny, ed., D. Reidel Publishing Co, Dordrecht-Holland (1979)
12. C. Braud, Comportement hydrophile-hydrophobe de copolymères d'acide acrylique et de N-(sec-butyl)-N-méthyl-acrylamide en solution aqueuse, Eur. Polym. J. 11:421 (1975)
13. C. Braud, Effect of hydrophobic interactions on the potentiometric behaviour of synthetic polyelectrolytes, in: "Colloques Internationaux du CNRS n°246: l'Eau et les Systèmes Biologiques," édition du CNRS, Paris (1976)
14. C. Villiers and C. Braud, Influence de l'ionisation sur le comportement conformationnel de copolymères alternés acide maléique/alkylvinyléther dans l'eau : effet de la structure du groupe alkyle, Nouv. J. Chim. 2:33 (1978)
15. M. Vert, F. Chabot, J. Leray and P. Christel, Stereoregular bioresorbable polyesters for orthopaedic surgery, Makrom. Chem., Suppl. 5:30 (1981)
16. M. Vert and R.W. Lenz, Malic acid polymers, US Patent n° 4,265,247, July 1 (1981)
17. V. A. Kropachev, Polymerization of heterocycles related to biomedical polymers, Pure Appl. Chem. 48:355 (1976)
18. S. S. Davis, Emulsion systems for the delivery of drugs by the parenteral route, in: "Optimization of Drug Delivery",

H. Bungaard, A. Bagger Hansen and H. Kofod, eds., Munksgaard, Copenhagen (1982)

19. C. Braud, M. Vert and E. Sélégny, Polyélectrolytes optiquement actifs à hydrophobie variable, Makrom. Chem. 175:775 (1974).

20. M. Mandel and M. G. Stadhouder, Conformational transition in partially esterified poly(methacrylic acid), Makrom. Chem. 80:141 (1964)

21. V. Bottiglione, Etude en mélange de solvants de l'acide polyméthacrylique et de copolymères acide méthacrylique-co-méthacrylate de benzyle : influence de la tacticité - transitions conformationnelles, Third cycle Thesis, Lille-France (1978)

22. J. Morcellet-Sauvage, Etude de dérivés polyméthacryliques portant l'alanine dans leur chaîne latérale : comportement polyélectrolytique, propriétés conformationnelles et activité optique, Dr-es-Sc. Thesis, Lille-France (1982)

23. P. L. Dubin and U.P. Strass, Hypercoiling in hydrophobic polyacids, in: "Charged and Reactive Polymers - Polyelectrolytes and their Applications," A. Rembaum and E. Sélégny, eds., D. Reidel Publishing Co, Dordrecht-Holland (1975)

24. C. Braud, M. Vert and R.W. Lenz, Polyelectrolytical properties of poly(β-malic acid and of its partially benzylated derivatives, in: "Abstracts of communications of the 27th IUPAC Symposium on Macromolecules, Strasbourg-France (1981)

25. M. Vert and R. W. Lenz, Preparation and properties of poly-β-malic acid : a functional polyester of potential biomedical importance, ACS Polym. Prept. 20(1):608 (1979)

26. J. Selb and Y. Gallot, Distinction entre phénomènes d'agregation et de micellisation présentés par des copolymères amphipathiques - cas des copolymères polystyrène/polyvinylpyridinium en milieu aqueux, Makrom. Chem. 181:809 (1980)

27. C. Braud, C. Bunel, H. Garreau and M. Vert, Evidence for the amphiphilic structure of partially hydrogenolyzed poly(β-malic acid benzyl ester), Polym. Bull. 9:198 (1983)

28. H. Bundgaard, A. Bagger Hansen and H. Kofod, "Optimization of Drug Delivery," Munksgaard, Copenhagen (1982)

29. A. T. Florence, Drug solubilization in surfactant systems, in: "Techniques of Solubilization of Drugs," S.H. Yalkowsky, ed., M. Dekker, New York (1981)

30. G. Barone, V. Crescenzi, A. M. Liquori and F. Quadrifoglio, Solubilization of polycyclic aromatic hydrocarbons in poly(methacrylic acid) aqueous solutions, J. Phys. Chem. 71:2341 (1967)

POLYPENTAPEPTIDE OF ELASTIN AS AN ELASTOMERIC BIOMATERIAL

D. W. Urry, S. A. Wood, R. D. Harris and K. U. Prasad

Laboratory of Molecular Biophysics
University of Alabama in Birmingham
School of Medicine
Birmingham, Alabama 35294

SUMMARY

The polypentapeptide of elastin, (L$\cdot$-Val^1-L$\cdot$Pro^2-Gly^3-L$\cdot$Val^4-Gly^5)$_n$, is synthesized with $\bar{n} > 200$. The purity of the synthesis is verified by carbon-13 nuclear magnetic resonance spectra. Coacervate concentrations are γ-irradiation cross-linked at 2,6,10,14, 18,26 and 34 MRADs to produce elastomeric bands. The dependence of the elastic modulus on cross-linking dose is demonstrated and thermoelasticity studies indicate a dominantly entropic elastomeric force. These results are discussed in connection with the property of the polypentapeptide to self-align into fibers, in connection with a proposed entropy source for the elasticity utilizing a class of β-spiral conformations and in terms of specific analogs which had been synthesized to test concepts of conformation and entropy source. Finally, the biocompatibility of the polypentapeptide is considered.

INTRODUCTION

The elastic fiber occurs in virtually every tissue and organ in the body[1] and it provides both a resistance and a restoring force to deformation. In the aortic arch and descending thoracic aorta there is twice as much elastin as collagen.[2] Apart from the fine microfibrillar coating,[3] the elastic fiber is comprised of a single protein. The precursor protein prior to cross-linking is called tropoelastin,[4,5] and has been shown to contain repeating peptide sequences.[1,6,7] The most pronounced of these repeating sequences and the most striking primary structural feature in a 70% sequenced protein is the polypentapeptide,

17

$(L \cdot Val^1 - L \cdot Pro^2 - Gly^3 - L \cdot Val^4 - Gly^5)_n$ where n has been shown to be 11+ in pig[1] and 13+ in chick[8] tropoelastins.

Previous chemical cross-linking of synthetic polypentapeptide polymers with n≥40 showed the polymer spontaneously to form fibers which could be observed in the light microscope without fixative or staining of any kind,[9] which could be observed in the scanning electron microscope to splay out into many fine fibrils and to recoalesce back into the fiber,[10] and which, when negatively stained with uranyl acetate-oxalic acid at pH 6.2 and observed by transmission electron microscopy, appeared to be comprised of parallel aligned filaments with about a 50 Å periodicity[11] as verified by optical diffraction of the electron micrograph.[12] This is demonstrated in Figure 1. Stress-strain studies of the weakly chemically cross-linked polypentapeptide, depending on the extent of hydration, showed the synthetic matrix to be capable of exhibiting the same elastic modulus as the elastic fiber.[10]

Here are presented results on a very high molecular weight synthesis with n>200, on γ-irradiation cross-linking of this polymer in a concentrated viscoelastic state, on the elastic modulus as a function irradiation dose, and on the thermoelasticity of the γ-irradiation cross-linked polymer.

METHODS AND MATERIALS

Synthesis: The synthesis of the polypentapeptide, $(VPGVG)_n$, was carried out starting from the monomer Boc-GVGVP-OBzl. This sequence was used earlier[13] as the pentachlorophenyl ester for polymerization to obtain the polypentapeptide in about 44% yield and with an average molecular weight of about 9,000. A modified approach of synthesizing this polymer is used here which gave the polypentapeptide in 64% yield with molecular weight of more than 100,000. Boc-GVG-OBzl was assembled starting from glycine benzyl ester by the mixed anhydride method using isobutyl chloroformate.[14] The tripeptide benzyl ester was hydrogenated and coupled with H-VP-OCH$_3$ obtained from Boc-VP-OCH[3] using EDCI[15]/HOBt. The pentapeptide Boc-GVGVP-OCH$_3$ was saponified and converted into the p-nitrophenyl ester using bis-(p-nitrophenyl) carbonate.[16] After removing the Boc group, the peptide p-nitrophenyl ester was polymerized in DMSO for fourteen days, diluted with water, tube dialysed first using a 3,500 dalton cut-off membrane and then using 50,000 dalton cut-off dialysis tubing (obtained from Spectrum Medical Industries, Inc., Los Angeles, CA, 90054) and lyophilyzed. The intermediate products were all characterized by tlc, elemental analysis and C-13 nuclear magnetic resonance spectroscopy.

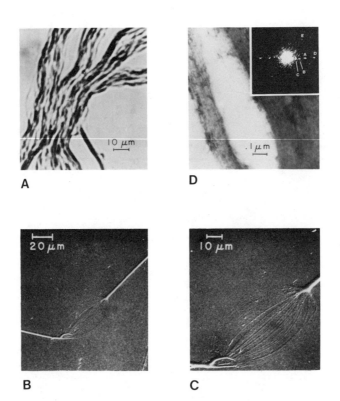

FIGURE 1:   Cross-linked polypentapeptide of elastin achieved with
polymers wherein the Val[4] residue on the average of one in every
fifth repeat was replaced by Lys in one synthesis and by Glu in a
second.   The two polymers were chemically cross-linked in water by
EDCI at 37°C in a water bath shaker.   A. Light micrograph in water
without fixatives showing the polymer to have self-aligned as
fibers. (reproduced with permission from reference 9).   B. and C.
Scanning electron micrograph of single self-assembled fiber seen
to splay-out into many fine fibrils and to recoalesce into the
same fiber (reproduced with permission from reference 10).   D.
Transmission electron micrograph of the same chemically cross-
linked polypentapeptide but negatively stained showing parallel
aligned filaments (adapted with permission from reference 11).
Insert.   Optical diffraction pattern of negatively stained
electron micrograph of polypentapeptide coacervate showing a major
spot A at about 50 Å (reproduced with permission from reference
12).   At each stage of observation, the polypentapeptide appears
to have self-assembled into fibers, comprised of fibrils,
comprised of parallel aligned filaments.

H - GVG - (VPGVG)$_n$ - VPV - OMe   in   Me$_2$SO-^{2}H$_6$

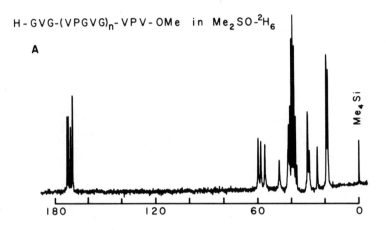

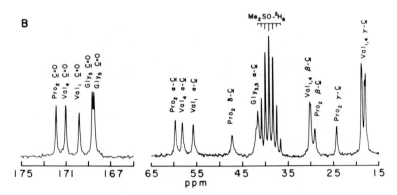

FIGURE 2:   Carbon-13 nuclear magnetic resonance spectrum of very
high molecular weight polypentapeptide of elastin. A.
Complete spectrum showing the absence of any extraneous
peaks; the multiple lines near 40 ppm are due to the
solvent. B. Expanded regions of the spectrum giving
the complete assignment of all resonances.

Abbreviations: Bis, N,N'-Methylene-bis-acrylamide; Boc,
tert·butyloxycarbonyl; Bzl - benzyl ester; ONp - p-nitrophenyl
ester; DMSO, dimethyl sulfoxide; EDCI, 1 (3-dimethylaminopropyl)3-
ethyl carbodiimide hydrochloride; HOBt, 1-hydroxybenzotriazole;
tlc - thin layer chromatography; PPP-polypentapeptide of elastin;
V, L·Val; P, L·Pro; G, Gly.

Nuclear Magnetic Resonance: The carbon-13 nuclear magnetic resonance spectrum shown in Figure 2 was obtained on a JEOL PFT-100P spectrometer at 25.15 MHz in dimethyl sulfoxide-d_6, at 23°C, 29 mg/0.5 ml with a 22 μsec pulse width, a 1 sec repetition time and proton noise decoupling. The spectrum in Figure 2A demonstrates the purity and correctness of the synthesis. The significant differences with previous data on the polypentapeptide with n≈18 are a splitting of the Gly^3 and Gly^5 carbonyl carbon resonances and greater splitting of the Val^1 and Val^4 γ-methyl resonances in the higher molecular weight polymer.

Coacervation: The polypentapeptide of elastin, as occurs with tropoelastin and fragmentation products of fibrous elastin, exhibits the interesting property of coacervation. These polypeptides are soluble in water below room temperature but on raising the temperature aggregation occurs and on standing at the elevated temperature a viscoelastic phase called the coacervate forms in the bottom of the vial. Temperature profiles of the aggregation process are given in Figure 3 for the polypentapeptide of elastin with $\bar{n}$>200 as a function of polymer concentration. This was after dialysis with a 3500 molecular weight membrane but before dialysis using a 50,000 molecular weight cut-off membrane. The coacervate phase at 40°C for the greater than 50,000 molecular weight fraction is about 50% peptide by weight. Additionally, the coefficient

of thermal expansion, $\beta = \left(\dfrac{\delta \ln V}{\delta T} \right)_P$ is approximately zero in the

40°C to 70°C temperature range.

Molecular Weight Determination: Molecular weight analysis of the >50,000 dialyzed fraction of $(GVGVP)_n$ was determined by SDS - polyacrylamide gel electrophoresis using the method of Swank and Munkres.[17] 100-200 μg of ^{14}C formylated PPP was loaded onto either 7.5% or 10% gels with a bis:acrylamide ratio of 1:30 cast in 6 mm I.D. X 130 mm tubing. Samples were electrophoresed at 5 mA per gel (roughly 2-3 volts/cm) for 4-5 hours in a Bio-Rad Model 150 gel electrophoresis cell used in conjunction with a Bio-Rad Model 500/200 power supply with bromophenol blue (Fisher Scientific Co.) as the tracking dye. The gels were removed and sliced into sequential sections with a Bio-Rad Model 190 gel fractionator, solubilized in 4 ml Ready-Solv MP liquid scintillation cocktail (Beckman Cat. No. #566763) and counted using the LKB Model 1217 liquid scintillation counter with corrections for quenching. A molecular weight determination of greater than 100,000 daltons was made by concurrent electrophoresis and comparison to 70 μg of a Pharmacia Inc. (#17-0446001) protein calibration kit stained with Coomassie Blue G-250 (Bio-Rad). As the pentamer molecular weight is 409, this gives a value of greater than 200 for $\bar{n}$.

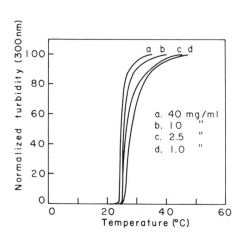

FIGURE 3: Temperature profile of aggregation followed as turbidity at 300 nm and given as a function of concentration for the very high molecular weight polypentapeptide of elastin. This synthesis with $\bar{n}>200$ gives
a high concentration limit for the onset of coacervation at 24°C. Lower molecular weight polymers have
curves that are shifted a few degrees to higher temperatures.

Irradiation Cross-linking: For irradiation cross-linking, 100
mg of polymer (n≳200) was dissolved in 100 λ H_2O in a polypropylene cryotube (Vangard International, Cat. No. 1076) fitted with a
machined Plexiglas pestle containing a 1 mm deep x 3.5 mm wide
circumferential channel. The samples were raised to coacervate
temperature, 40°C, shear-oriented, sealed to prevent drying,
quenched in liquid nitrogen and irradiated at the Auburn
University Nuclear Science Center as previously described.[18]
Exposure to a Cobalt-60 source at 13,317 roentgens per minute was

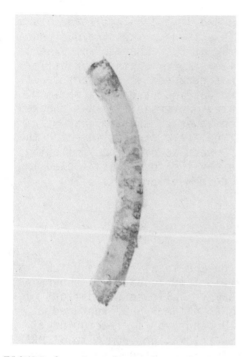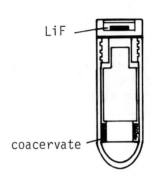

LiF

coacervate

FIGURE 4: Cross-section of the coacervation chamber configuration
showing the fitted pestle, the 1 x 3.5 mm circumferential channel, the LiF dosimeter and the polypropylene
tube which upon γ-irradiation cross-linking resulted in
the formation of elastomeric bands of PPP. One such
band is seen at the left.

for sufficient times to result in final γ-irradiation doses of
2,6,10,14,18,26 and 34 MRADS. LiF dosimetry[19] was used as a check
of radiation doses received by each sample. In all instances,
except for the 2 MRAD polymer sample, elastomeric bands were
formed upon cross-linking (see Figure 4).

Stress-Strain & Thermoelasticty Characterization: The
Young's modulus of the elastomeric polymer bands was determined
using force data collected at 22°C at a strain rate of 2 mm/min on
an apparatus designed and built in this laboratory.[18] Cross-
sectional areas were determined by measurement of sample width and
thickness before placement in the specimen grips. Initial lengths
of 5 mm were used and extensions of 40, 60, 80 and 100 percent
were recorded when possible.

For the thermoelastic characterization of the 6,10,14 & 18
MRAD cross-linked polymer bands, the sample was allowed to obtain
swelling equilibrium then stretched to 40% or 50% elongation at

23

either 22° or 60°C while being immersed in a constant temperature bath with a large excess of water as solvent. The strain elongation $\alpha = L/L_i$ was calculated using L_i at 22°C where L and L_i are the stretched and unstretched swollen sample lengths respectively. When elongated at 22°C, the force was recorded at constant length at 3-5° intervals as the temperature was increased to a maximum of 60°C at a rate of less than 1°/minute. Reversibility was checked by recording the force observed upon lowering the temperature at similar rates and decrements. Thermoelastic studies in which the polymer bands were stretched at 60°C were accomplished following the method of Mistrali et al.[20] whereby the force was allowed to equilibrate at each 5° temperature decrement prior to being recorded.

RESULTS

The effect of irradiation dose on the total force and subsequent rupture elongations of the elastomeric bands is shown in Figure 5A. The 6,10, and 14 MRAD samples allowed elongations of greater than 80-100% while the 18 MRAD polymer band ruptured at close to 60% with the 26 and 34 MRAD bands easily fracturing at roughly 40%. Although increasing γ-irradiation from 6 to 34 MRADS resulted in a ten fold increase in the total stress sustained by the elastomeric bands, the extent to which the sample could be elongated without failure due to rupture was reduced by roughly 1/2 to 2/3. This indication of possible increased backbone chain damage of the polymer matrix upon increasing irradiation levels may eventually restrict the extent to which this method of cross-linking can be utilized to obtain an elastomeric matrix similar to that of normal elastic tissue.

Concurrent with an increase in stress with increased irradiation dose is an increase in the Young's modulus of elasticity from 0.4×10^5 dynes/cm^2 at 6 MRADS to 13×10^5 dynes/cm^2 at 34 MRADS as seen in Figure 5B. Values reported for the 6,10,14 and 18 MRAD elastomeric bands are at 50% extention while the 26 and 34 MRAD values are at 38% and corrected for the cross-sectional area of the sample at the site of rupture.

The ratio of the internal energy component, f_e, of the force to the total elastomeric force, f_e/f, for the 6,10,14 and 18 MRAD samples is determined from the experimentally observed values of $\ln f/T$ as a function of temperature, following the method of Andrady and Mark.[21] Figure 6 illustrates the linearity, reversibility and near zero slopes of the thermoelastic data obtained for the irradiated samples after elongation to the indicated values of α at 22°C. The values of f_e/f observed for the cross-linked PPP bands in Figure 6 from the relationship

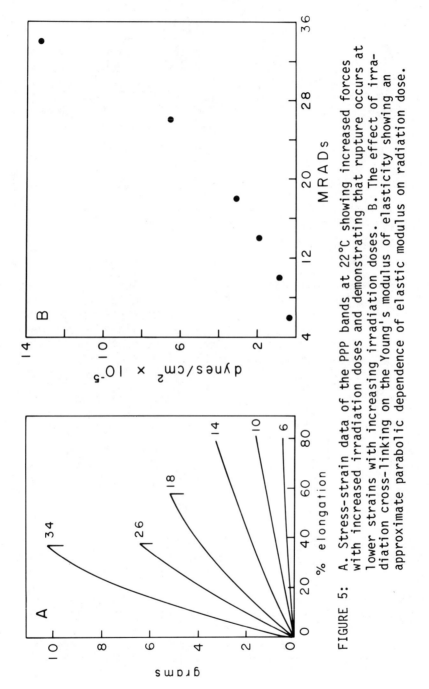

FIGURE 5: A. Stress-strain data of the PPP bands at 22°C showing increased forces with increased irradiation doses and demonstrating that rupture occurs at lower strains with increasing irradiation doses. B. The effect of irradiation cross-linking on the Young's modulus of elasticity showing an approximate parabolic dependence of elastic modulus on radiation dose.

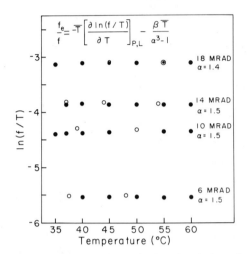

FIGURE 6: Thermoelasticity data obtained after elongation of the polymer bands at 22°C in H_2O. The solid circles represent experimental values upon heating with the open circles indicating the values obtained on cooling. The value of α is given with respect to the swollen sample initial length, L_i, at 22°C.

$$f_e/f = -T\left(\frac{\delta \ln (f/T)}{\delta T}\right)_{V,L}$$

were -0.09 (σ=0.25) and -0.05 (σ=0.23) using the slopes and average temperatures from the 45°-60° and the 50°-60°C ranges, respectively. The correction, for the contribution due to thermal expansion of the samples, was taken as zero due to a near zero thermal expansion coefficient for the PPP coacervate in the

26

40°-70° range. Mistrali et al.[20] similarly observed β=0 for elastin in H_2O in the ~50° to 70°C range. Both Mistrali et al.[20] and Andrady and Mark[21] have determined the corrections due to thermal expansion to be of the order of 0.10. The dependence of f_e/f on the degree of crosslinking is within the error of reproducibility. Mark[22] has reported that f_e/f values are commonly independent of extent of cross-linking although there are other studies[23] where a dependence has been reported.

DISCUSSION

The carbon-13 nuclear magnetic resonance spectrum of Figure 2 serves to verify the synthesis and adds the feature that the higher molecular weight polymer results in removal of the overlap of the Gly[3] and Gly[5] carbonyl carbon resonances and also removes overlap of the Val[1] and Val[4] side chain resonances. The temperature profiles for aggregation (see Figure 3) are both concentration and chain length dependent with the present synthesis giving the lowest temperature onset for aggregation.[24] Another feature consistent with the larger molecular weight of the present synthesis is that the coacervate is opaque whereas for lower molecular weight polymers it is transparent.

The present results demonstrate that the elastic modulus is proportional to the γ-irradiation dose in a near parabolic manner, but the higher γ-irradiation doses lead to rupture of the elastomeric bands at lower extensions. While higher values for the elastic modulus need to be obtained in order to provide more accurate values of the f_e/f ratio, it does appear possible to conclude that the elastomeric force is dominantly entropic in origin. This raises the interesting question as to how a polymer system that is sufficiently anisotropic to self-align in the formation of fibers, fibrils and filaments, as shown in Figure 1, can be an entropic elastomer. This issue has been previously discussed[9,24,25] in terms of a proposed librational entropy mechanism of elasticity arising from a class of β-spiral (helical) conformations. The proposed mechanism has been tested by selected analogs: the L·Ala[5] PPP,[26] the D·Ala[5] PPP[27] and the D·Ala[3] PPP.[28] The addition of methyl residues at the α-carbon of Gly[5] destroys elasticity. In the case of the L·Ala[5] PPP while presenting a temperature profile of aggregation essentially identical to that of Figure 3 results in a granular precipitate with no elastic properties.[26] In the case of the D·Ala[5] PPP aggregation to form a coacervate phase is extremely slow taking days to form but on cross-linking the matrix simply fragments under stress.[27] The D·Ala[3] PPP on the other hand readily forms a viscoelastic coacervate phase which when cross-linked by γ-irradiation results in a higher elastic modulus at a lower radiation dose.[28]

With insight typical of his many research contributions P. J. Flory paraphrased a basic postulate of rubber elasticity as "In other words, the stress exhibited by a strained specimen of rubber is asserted to be intramolecular in origin; intermolecular forces between chains in contact are assumed to be inconsequential insofar as the stress is concerned."[29]   It appears, in the hundreds of millions of years available for evolution of a suitable biological elastomer capable of functioning in an aqueous milieu for the lifetime of an organism, that a molecular construct was developed which adheres to this postulate and utilizes a conformational construct based on a faithfully retained repeating peptide sequence. The same polypeptide repeat is accurately retained in species as diverse as chick and pig.[5,8]   One of a class of the proposed molecular constructs is shown in Figure 7 and it is thought that two or three of these β-spirals supercoil to form the 50 Å filaments observed in the electron microscope.[25] Viewing each β-spiral as a supercoiled cylinder, the only interchain contacts are the weak hydrophobic contacts between Val and Pro side chains. Even the contacts between turns of the dynamic helices are hydrophobic with the Type II Pro^2-Gly^3 β-turns functioning as spacers between the turns of the β-spiral and on slight extension the β-turns function to suspend the Val^4-Gly^5-Val^1 segment allowing this segment to provide the primary source of entropy. This is why the L·Ala^5 and D·Ala^5 analogs were synthesized, to test this site as an essential element of the motion giving rise to entropy. The D·Ala^3 analog underscores the Type II β-turn as a central conformational feature. A D·residue in position-3 of a Type II β-turn is known to stabilize this conformational feature.[30]   It may be noted that in a survey of 29 crystal structures the Pro-Gly sequence is the most probable β-turn which is a conformational feature more probable than the β-pleated sheet and almost as probable as the α-helix.[31]   It may also be noted that the β-spiral of Figure 1 was derived[32] after the cyclopentadecapeptide had been identified as the cyclic conformational correlate of the linear polypentapeptide[33] and after solution[34] and crystal conformations[35] of the cyclic correlate had been determined.

When considering the polypentapeptide of elastin as a potential biomaterial, there are two aspects that should be noted; one is its propensity to calcify and the second concerns the issue of biocompatibility. When weakly cross-linked and flow oriented, the PPP of elastin has an inherent tendency to calcify[36] when exposed to normal bovine serum as does the elastic fiber itself.[37] However, as cross-linking is increased calcification decreases. Being a natural repeating peptide sequence, it is not surprising that the polypentapeptide has little or no effect on human gingival fibroblasts in cell culture as measured in terms of cell viability[38] and cell function followed by means of reproduction

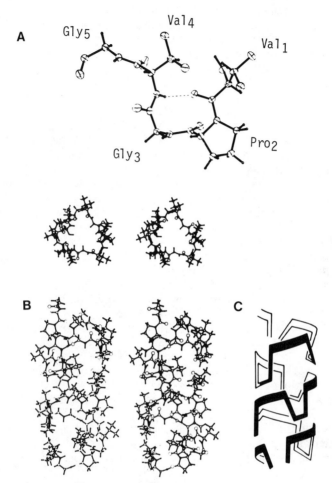

FIGURE 7: A. Conformation of the repeating unit of the cyclic
conformational correlate, cyclo (VPGVG)$_3$, of the linear
polypentapeptide, obtained by X-ray diffraction and
shown to be nearly identical to the solution conformation. (Reproduced with permission from reference
35.) B. β-spiral generated by going linear from the
cyclic conformational correlate. Above is a spiral
axis view. Note the space for water within the β-
spiral, the interturn contacts of weak hydrophobic
interactions, the suspended Val^4-Gly^5-Val^1 segment and
the β-turns functioning in the relaxed state as spacers
between turns of the spiral. (Reproduced with permission from reference 32). C. Schematic representation showing the β-turns functioning as spacers
between turns of the β-spiral and showing gaps in the
surface where bulk water is contiguous with intraspiral
water (Reproduced with permission from reference 9).

and protein synthesis.[39] Additionally in rabbit implant studies at three weeks histological examination showed normal tissue distribution with no foreign body response and at thirteen weeks by radiographic analysis there was no evidence of inflammation.[40]

ACKNOWLEDGEMENT

This work was supported in part by the National Institutes of Health Grant No. HL-29578 and United States Army Contract No. DAMD17-82-C-2129. The authors wish to thank T. L. Trapane for obtaining the spectrum of Figure 2.

REFERENCES

1. L. B. Sandberg, N. T. Soskel and J. B. Leslie, Elastin structure, biosynthesis and relation to disease states. N. Engl. J. Med. 304:566 (1981).
2. G. M. Fischer and J. G. Llaurado, Collagen and elastin content in canine arteries selected from functionally different vascular beds. Circ. Res. 19:394 (1966).
3. R. Ross and P. Bornstein, The elastic fiber. I. The separation and partial characterization of its macromolecular components. J. Cell Biol. 40:366 (1969).
4. D. W. Smith, N. Weissman and W. H. Carnes, Cardiovascular studies on copper deficient swine. XII. Partial purification of a soluble protein resembling elastin. Biochem. Biophys. Res. Commun. 31:309 (1968).
5. L. B. Sandberg, N. Weissman and D. W. Smith, The purification and partial characterization of a soluble elastin-like protein from copper-deficient porcine aorta. Biochemistry 8:2940 (1969).
6. W. R. Gray, L. B. Sandberg and J. A. Foster, Molecular model for elastin structure and function. Nature 246:461 (1973).
7. J. A. Foster, E. Bruenger, W. R. Gray and L. B. Sandberg. Isolation and amino acid sequences of tropoelastin peptides. J. Biol. Chem. 248:2876 (1973).
8. L. B. Sandberg (private communication).
9. D. W. Urry, What is elastin, what is not. Ultrastruct. Pathol. 4:227 (1983).
10. D. W. Urry, K. Okamoto, R. D. Harris, C. F. Hendrix and M. M. Long, Synthetic, cross-linked polypentapeptide of tropoelastin: An anisotropic, fibrillar elastomer. Biochemistry 15:4083 (1976).
11. D. W. Urry and M. M. Long, On the conformation, coacervation and function of polymeric models of elastin. in: "Elastin and Elastic Tissue," L. B. Sandberg, W. R. Gray and C. Franzblau, eds., Plenum Press, New York, Adv. Exp. Med. Biol., 79:685-714, (1977).

12. D. Volpin, D. W. Urry, I. Pasquali-Ronchetti and L. Gotte, Studies by electron microscopy on the structure of coacervates of synthetic polypeptides of tropoelastin. Micron 7:193 (1976).
13. J. R. Bell, R. C. Boohan, J. H. Jones and R. M. Moore, Sequential polypeptides. Part IX. The synthesis of two sequential polypeptide elastin models. Int. J. Peptide and Protein Res. 7:229 (1975).
14. J. R. Vaughan, Jr. and R. L. Osato, The preparation of peptides using mixed carbonic-carboxylic acid anhydrides. J. Am. Chem. Soc. 74:676 (1952).
15. J. C. Sheehan, J. Preston and P. A. Cruickshank, A rapid synthesis of oligopeptide derivatives without isolation of intermediates. J. Am. Chem. Soc. 87:2492 (1965).
16. Th. Wieland, B. Heinke and K. Vogeler, Derivate der kohlensäure in der peptidchemie. Ann. 655:189 (1962).
17. R. T. Swank and K. D. Munkres, Molecular weight analysis of oligopeptides by electrophoresis in polyacrylamide gel with sodium dodecyl sulfate. Anal. Biochem. 39:462 (1971).
18. D. W. Urry, R. D. Harris and M. M. Long, Irradiation cross-linking of the polytetrapeptide of elastin and compounding to dacron to produce a potential prosthetic material with elasticity and strength. J. Biomed. Mater. Res. 16:11 (1982).
19. W. L. McLaughlin, A. C. Lucas, B. M. Kaspar and A. Miller, Electron and gamma-ray dosimetry using radiation-induced color centers in LiF. in: Trans. 2nd International Meeting on Radiation Processing, J. Silverman, ed., 2-5, Miami, Florida, (1978).
20. F. Mistrali, D. Volpin, G. B. Garibaldo and A. Ciferri, Thermodynamics of elasticity in open systems. Elastin. J. Phys. Chem. 75:142 (1971).
21. A. L. Andrady and J. E. Mark, Thermoelasticity of swollen elastin networks at constant composition. Biopolymers 19:849 (1980).
22. J. E. Mark, Thermoelastic properties of rubberlike networks and their thermodynamic and molecular interpretation. Rubber Chem. Technol. 46:593 (1973).
23. J. E. Mark, Thermoelastic results on rubberlike networks and their bearing on the foundations of elasticity theory. J. Polymer Sci., Macromolecular Reviews 11:135 (1976).
24. D. W. Urry, Characterization of soluble peptides of elastin by physical techniques. in: Methods in Enzymology, L. W. Cunningham and D. W. Frederiksen, eds., Academic Press, Inc., New York, 82:673, (1982).
25. D. W. Urry, C. M. Venkatachalam, M. M. Long and K. U. Prasad, Dynamic β-spirals and a librational entropy mechanism of elasticity. in: Conformation in Biology, R. Srinivasan and R. H. Sarma, eds., G. N. Ramachandran Festschrift Volume, Adenine Press, USA, 11-27, (1982).
26. D. W. Urry, T. L. Trapane, M. M. Long and K. U. Prasad, J. Chem. Soc., Faraday Trans. I, 79:853, 1983.

27. D. W. Urry, T. L. Trapane, S. A. Wood, J. T. Walker, R. D. Harris and K. U. Prasad, Int. J. Pept. Protein Res. 22:164 (1983).

28. D. W. Urry, T. L. Trapane, S. A. Wood, R. D. Harris, J. T. Walker and K. U. Prasad, Int. J. Pept. Protein Res. (in press).

29. P. J. Flory, Molecular interpretation of rubber elasticity. Rubber Chem. Technol. 44:G41 (1968).

30. C. M. Venkatachalam, Stereochemical criteria for polypeptides and proteins - Part V. Conformation of a system of three linked peptide units. Biopolymers 6:1425 (1968).

31. P. Y. Chou and G. Fasman, β-turns in proteins. J. Mol. Biol. 115:135 (1977).

32. C. M. Venkatachalam and D. W. Urry, The development of a linear helical conformation from its cyclic correlate. The β-spiral model of the elastin polypentapeptide, $(VPGVG)_n$. Macromolecules 14:1225 (1981).

33. D. W. Urry, T. L. Trapane, H. Sugano and K. U. Prasad, Sequential polypeptides of elastin: cyclic conformational correlates of the linear polypentapeptide. J. Am. Chem. Soc. 103:2080 (1981).

34. C. M. Venkatachalam, M. A. Khaled, H. Sugano and D. W. Urry, Nuclear magnetic resonance and conformational energy calculations of repeat peptides of elastin: conformational characterization of cyclopentadecapeptide, cyclo-$(L-Val_1-L-Pro_2-Gly_3-L-Val_4-Gly_5)_3$. J. Am. Chem. Soc. 103:2372 (1981).

35. W. J. Cook, H. M. Einspahr, T. L. Trapane, D. W. Urry and C. E. Bugg, The crystal structure and conformation of the cyclic trimer of a repeat pentapeptide of elastin, cyclo-$(L-Valyl-L-Prolyl-Glycyl-L-Valyl-Glycyl)_3$. J. Am. Chem. Soc. 102:5502 (1980).

36. D. W. Urry, M. M. Long, C. F. Hendrix and K. Okamoto, Cross-linked polypentapeptide of tropoelastin: An insoluble, serum calcifiable matrix. Biochemistry 15:4089 (1976).

37. D. W. Urry, Molecular perspectives of vascular wall structure and disease: The elastic component. Perspect. Biol. Med. 21:265 (1978).

38. A. Stevens, R. Cogen, D. W. Urry and M. Long, Effect of a calcifiable matrix on human cell viability. J. Dental Research 60:391 (1981).

39. A. Waikakul, R. Cogen, A. Stevens, D. Urry and M. Long, Effect of a calcifiable matrix on human cells. J. Dental Research 61:189 (1982).

40. J. Lemons, personal communications.

DEVELOPMENT OF NON-THROMBOGENIC MATERIALS

J. L. Williams, A. Rumaks, J. P. O'Connell

Becton Dickinson Research Center
P. O. Box 11016
Research Triangle Park, North Carolina 27709

INTRODUCTION

The development of non-thrombogenic materials and surface treatments are ongoing needs in the biomaterials area. Also, the need for improved test procedures for evaluating new materials in terms of thrombogenicity continues to be a requirement.[1,2] In the current context, results of surface treatments for improved non-thrombogenicity along with a description of the test method employed will be discussed. Also, results will be presented to substantiate the validity of the Dudley Blood Clotting[3] for directly measuring surface thrombogenicity.

EXPERIMENTAL

The method employs materials in the form of small diameter tubing or to materials that can be adequately coated onto the inside of small diameter tubing. In general, a 12-inch length of tubing with an inner diameter of 0.020 inch is preferred. This yields a tube with a volume of 60 µl, with a surface/volume ratio of about 80 cm^2/ml blood.

After the dog has been anesthetized and the neck cleaned and disinfected, the jugular vein is located. Without injury to the vessel, a puncture wound is made through the skin and underlying tissue with a standard 14-gauge needle. The catheter placement set is then inserted through the wound and into the jugular vein; the needle is removed from the catheter. The sample tubing is inserted through the catheter and, while slowly infusing 1 to 2 ml of sterile normal saline solution through the test tubing,

33

advanced into the vein up to the mark. The sample tube is now in the jugular vein and extends about 0.5 cm beyond the tip of the indwelling catheter. The other end of the sample tube is immersed in a small flask of corn oil as shown in Figure 1 to prevent air-interface clotting. The corn oil forces the blood to form discrete drops as it leaves the tubing, thus providing a simple method of monitoring flow; gravimetric or volume displacement methods can also be used. A timer is started as the first drop of blood leaves the tubing and is stopped when the flow of blood stops. The elapsed time is the Dudley Clotting time. The clotted sample tube is removed, and the indwelling Teflon catheter gently flushed with sterile saline solution. A second sample tube can then be inserted, and the procedure repeated. When the last sample has been tested, the indwelling catheter is removed, and pressure applied to the wound until the bleeding stops. Depending on the amount of blood lost, the dog is rested for a minimum of one week before using again. The flow rate through 0.020 inch tubing is about 10 ml per hour, so that blood loss is minimal.

Heparinized materials were prepared according to the details given in the Dudley-Williams reference.[4]

RESULTS AND DISCUSSION

Several polymeric materials were examined using the Dudley Clotting Test described herein. The results of those tests are summarized in Table 1 below.

Table 1. Polymeric Materials Evaluated by Dudley Test

Material	Number Tested	Average Clotting Time (minutes)	± s.d.
Polyurethane [1]	40	18.9	3.3
Polyethylene	12	13.9	2.0
Teflon	9	10.1	1.4
Silicone Rubber	20	8.2	1.7
Stainless Steel	5	6.8	1.7
Acrylic Acid/ Polyethylene [2]	5	6.5	0.8
Hema/Polyurethane [2]	5	9.2	1.3
PVP/Polyurethane [2]	5	11.9	1.0

[1] Upjohn Pellethane
[2] Irradiation grafted

34

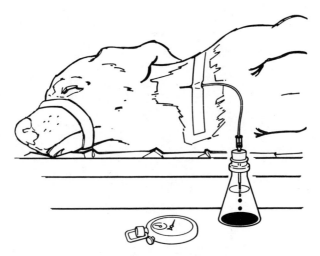

Fig. 1. *Ex Vivo* Dudley Clotting Test.

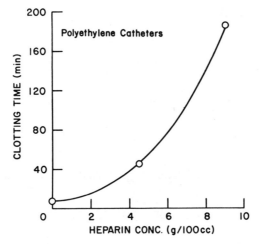

Fig. 2. Effect of heparin bath concentration on Dudley Clotting
times with polyethylene.

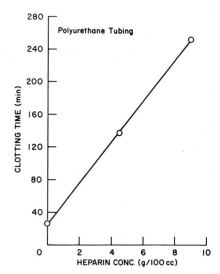

Fig. 3. Effect of heparin bath concentration on Dudley Clotting
times with polyurethane.

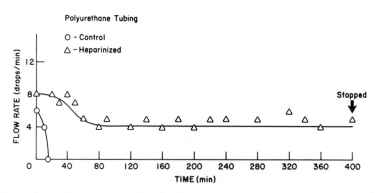

Fig. 4. Flow rate of blood in Dudley Clotting Test for
polyurethane before and after heparinization.

Clearly, polyurethane is less non-thrombogenic than the remaining materials summarized in Table 1. Also, stainless steel which is commonly encountered in needles appeared the most thrombogenic in this study.

Two materials, polyethylene and polyurethane, were selected for surface modification using heparinization techniques developed in the laboratory.[4] The results for heparinized polyethylene and polyurethane are shown in Figures 2 and 3, respectively. In both cases, an increase in the concentration of sodium heparin used in the second step of the formation of the heparin complex, i.e.,

$$RCH_3^{\oplus} Cl^{\ominus} + Na^{\oplus} Heparin^{\ominus} \rightarrow RCH_3^{\oplus} Heparin^{\ominus} + Na^{\oplus} Cl^{\ominus}$$

resulted in an increase in blood clotting times. However, the best overall increase was found in the case of polyurethane which is more receptive to the quaternary salt used in the first step of the process.

The kinetics of a more refined heparinized polyurethane are shown in Figure 4 before and after heparinization. In this case, the untreated polyurethane tube had clotted after 19 minutes while the heparinized material remained clear even after seven hours.

CONCLUSIONS

These experiments demonstrate the simplicity and utility of the Dudley Clotting Test for evaluating blood thrombogenicity of foreign surfaces. Also, the results point to the need of a heparinized surface for prolonged non-thrombogenicity when polymeric materials are in contact with blood for extended periods of time.

REFERENCES

1. P.M. Galletti, J.L. Brash, K.H. Keller, G. La Forge, R.G. Mason, W.S. Pierce, and J.A. Reynolds, "Report of the Task Force on Biomaterials to the Cardiology Advisory Committee of the NHLBI," Artificial Organs, 2, 189-201 (1978).

2. A. Achorya, J.F. Brown, W.J. Dodds, S. Hanson, L. Hasner, R.H. Kahn, E. Nyilas, P.D. Richardson, E.W. Salzman, J.T. Watson, R. Wineman, F.A. Pittick, R.G. Mason, "Working Group on Blood-Material Interactions," Final Report to the Devices and Technology Branch, NHLBI (1979).

3. B. Dudley, J.L. Williams, K. Abel, and B. Muller, "Synthesis and Characterization of Blood Compatible Surfaces, Part I. Dynamic Tube Test Applied to Heparinized Surfaces," Trans. Amer. Soc. Certif. Int. Organs, 22, 538-543 (1976).

4. U.S. Patent 4,116,898, January 12, 1979.

THE USE OF POLYACRYLATES IN THE MICROENCAPSULATION OF

MAMMALIAN CELLS

Francis V. Lamberti, Ramon Evangelista, Margaret A.
Wheatley, John Blysniuk and Michael V. Sefton

Department of Chemical Engineering and Applied
Chemistry, University of Toronto, Toronto, Ontario
M5S 1A4

Diabetes is a metabolic syndrome that affects one in twenty persons. Almost one half of these require repeated, often daily insulin injections to maintain approximately normal glucose levels. Nevertheless, the long-term degenerative sequelae of diabetes -- blindness, heart disease and kidney failure -- make it the third leading cause of death in North America.

A direct causal relationship between the degree of diabetic control and the development of diabetes-related degenerative complications has yet to be unambiguously demonstrated. However, the growing evidence that the long term complications of diabetes can be minimized by better control of hyperglycemia has prompted the development of a wide range of continuous insulin infusion systems[1-3] designed to provide either closed- or open-loop control of glycemia. Alternatively, physiological control of insulin-dependent diabetes, may be achieved by the transplantation of pancreatic islets. To prevent their rejection by the host's antibodies, it has been suggested[4] that these cells be immobilized in a biocompatible semipermeable tissue compatible matrix.

Mammalian tissue cell immobilization or microencapsulation has been achieved by processes involving either interfacial precipitation[5] or interfacial adsorption[4,6]. Interfacial precipitation methods are restricted to water-insoluble materials, while interfacial adsorption of capsule wall forming materials to a semi-solid matrix containing cells has been successfully applied to mammalian cell encapsulation using both water soluble and water-insoluble materials[6].

Calcium alginate immobilized pancreatic islets coated with polylysine have been demonstrated to successfully ameliorate streptozotocin induced diabetes in rats for up to two months[7]. However, these capsules were not sufficiently chemically or mechanically durable for long term implants. The slow dissolution

39

of the water-soluble components of the capsule wall may have triggered a chronic inflammation of the surrounding tissue to produce the observed influx of macrophages. Alternatively, the inflammatory response may have been elicited by the release of the capsule contents due to mechanical stress.

Rather than exploring the wide range of capsule wall forming materials available, we have focussed on the application of hydrophilic, water-insoluble polyacrylates to the production of durable, implantable microencapsulated tissue cells. Model tissue cells (human erythrocytes) have been employed to assess the physical and metabolic consequences of mammalian tissue cell encapsulation. Two methods have been investigated: interfacial precipitation from an organic solution and interfacial adsorption from a stable aqueous emulsion onto a polysaccharide based hydrocolloid cell immobilization matrix.

MATERIALS AND METHODS

Polymers

EUDRAGIT RL100 (Röhm Pharma, Darmstadt, W. Germany) an acrylic methacrylic acid copolymer with a low content of quaternary ammonium groups (alkali value of 29 mg KOH/g dry lacquer) was used either as a 12.0% (w/v) solution in diethyl phthalate (DEP) or as an emulsion in isotonic tris-buffered saline containing 10 mM $CaCl_2$ (TBC). The emulsion was prepared by dissolving 5% (w/v) EUDRAGIT in boiling, glass distilled water, cooling and diluting to the desired concentration.

A poly (HEMA-MMA) copolymer was synthesized from an equimolar mixture of the monomers by solution polymerization to a low degree of polymerization using an exisiting method[8]. Briefly, the monomers, 2-hydroxylethyl methacrylate (HEMA) (Monomer-Polymer Lab; Borden Chemical Co., Philadelphia, PA) and methyl methacrylate (MMA) (Poysciences, Warrington, PA), were distilled in vacuo prior to polymerization. The monomers were added to ethanol and refluxed at 75°C. Polymerization was initiated by the addition of azobisisobutyronitrile (Eastman Kodak, Rochester, NY). After 4 hours, polymerization was terminated by the addition of excess solvent. The reaction mixture was quenched in an ice bath and poured off into hexane to precipitate the polymer. The product was redissolved in acetone, precipitated in water and dried under vacuum. The equilibrium water content of the product swollen in distilled water was 15% (w/w).

Sodium alginate (Kelcogel LV, Kelco, Chicago, IL), a soluble polysaccharide isolated from brown algae and composed of β-D mannuronate and α-L-guluronate[9], was used without purification. In the presence of calcium or other metal salts (except magnesium), alginate forms an ionically cross-linked gel. Alginate immobilization has been demonstrated to be an effective yet gentle process which does not affect cell viability[4,10].

Cells

Human erythrocytes were collected from healthy donors (Canadian Red Cross, Toronto, Ont.) and were stored for no more than 3 weeks in an anticoagulant citrate buffer at 4°C. Erythrocytes were prepared for encapsulation by centrifugation at 1470xg for two minutes (IEC Clinical Centrifuge, Needham Hts., MA), followed by three washes in an isotonic phosphate buffered saline (PBS) (containing in mM/L; 137NaCl, 1.47 KH_2PO_4 and 8.1 Na_2HPO_4; pH 7.4). Cells were suspended in PBS at a volume fraction of 0.5.

Pancreatic islets were obtained from a collagenase (Sigma Type V; Sigma Chemical Co., St. Louis, MO), digestion of pancreatic tissue prepared by a previously described method[12]. Individual islets were hand-picked under a dissecting microscope transferred to culture medium (CRML 1969, 13) and incubated at 37°C. Islets were prepared for encapsulation in EUDRAGIT by transfer to a small volume (2 mL media/1000 islets) of culture medium containing 10% (w/v) Ficoll 400 (Pharmacia, Uppsala, Sweden), as a stabilizer to prevent islet aggregation during processing.

Interfacial precipitation

An earlier microencapsulation process[5] using EUDRAGIT has been modified to reduce the shear stress on the cells during encapsulation and to prepare larger, more uniform capsules.

Droplets of cells surrounded by a solution of EUDRAGIT RL in DEP were produced by pumping the two liquids through coaxial needles, and allowing the liquids to mix briefly at the needle tip (Figure 1). The droplets were blown from the needle tip by a parallel air stream into a gently stirred corn oil-mineral oil mixture (70 v/v corn oil) precipitating the polymer solution around the aqueous core to form a nascent capsule wall which was hardened gradually by gentle mixing. The hardened capsules were collected by centrifugation or sieving, rinsed extensively in a 10% (w/v) bovine serum albumin solution in PBS to removal residual organic solvents and resuspended in PBS.

Erythrocytes were encapsulated in poly (HEMA-MMA) by a single needle variation of this procedure. A 10% (w/v) copolymer solution in polyethylene glycol 600 ((PEG), Polysciences, Warrington, PA) was mixed with washed erythrocytes to yield cells suspended in a homogeneous mixture consisting of copolymer, PEG and water (6.7:60:33.3 w/v/v respectively). The ratio of copolymer solution to cell suspension was selected to avoid premature precipitation of the copolymer. The mixture was extruded through a single needle as droplets into DEP (a partial non-solvent) to precipitate the polymer. Capsules were hardened by the slow diffusion of the copolymer solvent (PEG) into the DEP, followed by addition of PBS or a PEG solution in PBS. After centrifugation capsules were recovered from the DEP-water interface, and collected by stepwise washings in solutions of decreasing PEG concentration.

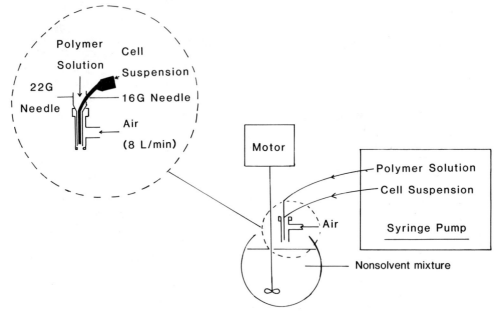

Fig. 1. Apparatus used for coextrusion of a capsule wall forming
material solution (outer needle) and a cell suspension
(inner needle). Droplets detached from the needle tip by
a coaxial air stream (air jacket) fall into a solvent-
nonsolvent mixture to form a nascent capsule wall
surrounding a core of an aqueous cell suspension.

Interfacial Adsorption

An aliquot of 0.1 mL freshly washed human erythrocytes was
mixed with 1.0 mL of a 2.0% (w/v) sodium alginate solution in
isotonic saline (0.15 M NaCl, 1 mM HEPES, pH 7.4) which had been
equilibrated at 37°C.   The cell suspension was extruded at 0.7
mL/min and the droplets were blown from the needle tip by a
coaxial air stream into an isotonic saline buffer containing 10 mM
$CaCl_2$ (TBC).   The resulting gel beads were allowed to harden and
were rinsed in a fresh curing bath for thirty minutes.

The excess liquid was removed by aspiration and the gel
immobilized cells were suspended in TBC containing 0.5% (w/v)
EUDRAGIT as a stable aqueous emulsion.   The beads were gently
shaken for up to two hours and were subsequently rinsed in fresh
buffer to remove unadsorbed polymer.

Qualitative Analysis of Microencapsulated Cells

A rapid qualitative assessment of cell morphology during
various stages of microencapsulation was obtained by examining
freshly prepared microcapsules using a transmitted light
microscope (Lietz SM Lux with Polaroid CB100 attachment, Wild
Lietz, Ottawa, Ont.).

Microencapsulated tissue cells were fixed and embedded using standard methods[14] at the Histology Research Laboratories of the Hospital for Sick Children (Toronto) by Mr. B. Wilson. Sections were stained using Harris's hematoxylin-eosin stain and examined by light microscopy.

Structural details of microcapsules were obtained by scanning electron microscopy of freeze-fractured samples. Briefly, microencapsulated tissue cells were fixed in a 1.0% (w/v) glutaraldehyde solution in PBS or TBC and transferred to stainless steel specimen studs. Excess liquid was removed, the samples were quench-frozen in liquid Freon and transferred to a specimen block under liquid nitrogen. Samples were fractured with a single blow from a chilled razor blade and were transferred under liquid nitrogen to a lyophilizer (Freeze-Dry 5, Labconco, Kansas City, MO).

Dried samples were mounted in specimen discs with an amalgam of oven-dried activated charcoal and an isocyanoacrylate adhesive. The samples were subsequently sputter-coated with a thin (10 nm) metallic film (Polaron Equipment Ltd., SEM coating unit with a Polaron Film Thickness Monitor E100; International Scientific Instruments, Santa Clara, CA) and examined (ISI-60 Scanning Electron Microscope, International Scientific Instruments Ltd.).

In Vivo Biocompatility Assay

Implantation of EUDRAGIT microcapsules was performed in cooperation with the Connaught Research Institute of Connaught Laboratories, Toronto, with the assistance of Ms. H. van der Rooy. Four male Wistar rats (250-350 g) were etherized and prepared for surgery. A small abdominal incision less than 1 cm in length was made below the sternum and the wound was opened to reveal the intraperitoneal cavity. A suspension of EUDRAGIT coated alginate microcapules (0.26 ± 0.02 g) sterilized by γ-irradiation (2.0 Mrads from a ^{60}Co source), was pipetted directly into the intraperitoneal cavity and the wound was closed. After 12 days, the rats were sacrificed and the abdominal cavity examined for gross tissue abnormalities. Tissue samples were removed from the implantation site and immediately transferred to Bouin's fixative. Histological preparation and staining of the excised tissue was performed by Dr. M. Wiley of the Anatomy Laboratories of the Faculty of Medicine at the University of Toronto.

RESULTS AND DISCUSSION

Interfacial Precipitation

Erythrocytes encapsulated in EUDRAGIT by precipitation from diethyl phthalate appeared to remain intact after encapsulation. Capsules produced by coextrusion of separate core and wall liquid phases were approximately 300 μm in diameter and consisted of a cellular core surrounded by a thin (~ 10 μm) polymeric shell (Figure 2). The size of the microcapsules was controlled by the flowrate of the air stream used to detach droplets from the needle

43

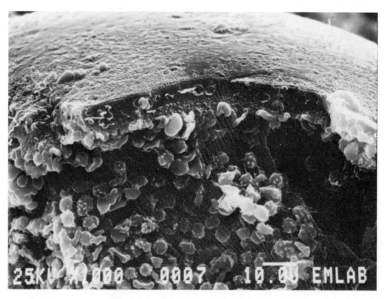

Fig.2.   SEM of freeze-fractured EUDRAGIT microencapsulated erythrocytes formed by coextrusion. The capsules were ~300 μm
in diameter with a 10 μm capsule wall.

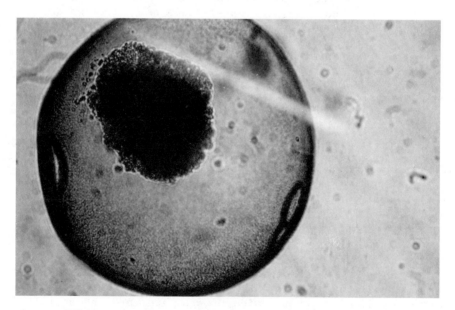

Fig.3.   Light micrograph of EUDRAGIT microencapsulated rat
pancreatic islets form by coextrusion. The capsules were
~300 μm in diameter, with a 10 μm capsule wall.

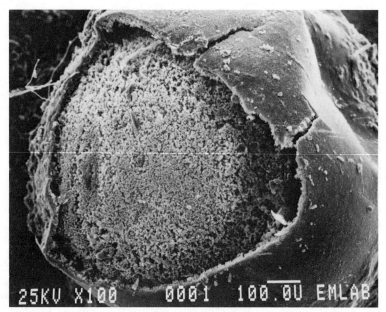

Fig.4. SEM of freeze-fractured poly(HEMA-MMA) microencapsulated
        erythrocytes formed by extrusion of a homogeneous mixture
        (6.7:60:33.3 (v/v) p(HEMA-MMA):PEG:RBC) into diethyl
        phthalate.

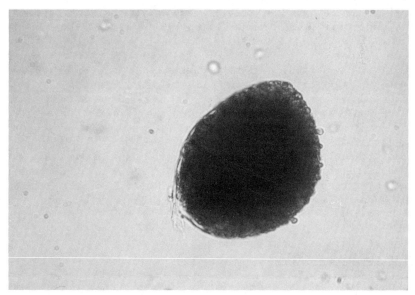

Fig.5. Light micrograph of EUDRAGIT coated calcium alginate gel
        containing immobilized erythrocytes. The capsules were
        ~1.9 mm diameter, with a 30 μm capsule wall (from ref.16).

tip while the thickness of the capsule wall depended on the relative flow rates of the two liquid phases and the polymer concentration.

Preliminary experiments with rat pancreatic islets have demonstrated that functionally differentiated mammalian tissue cells with active metabolisms may be encapsulated in EUDRAGIT by interfacial precipitation using the coextrusion method. No gross morphological changes were evident in microscopic examination (Figure 3) but final conclusions must await the quantitative assessment of tissue viability. Previous attempts at encapsulating these cells with EUDRAGIT by dispersion of the cells within the polymer solution and then adding a nonsolvent to effect precipitatin as was done for erythrocytes[5] was unsuccessful. Only erythrocytes were able to withstand the high shear stresses associated with dispersing the cells by magnetic or overhead stirring in the polymer solution. Although intact islets were recovered after dispersion alone, the combination of dispersion and interfacial precipitation caused severe disruption of the islet structure. The coextrusion process minimizes the mechanical damage done to the cells, and allowed better control of the polymer-to-cell ratio so that small quantities of cells (e.g., ~ 100 islets) could easily be handled. Without this control it proved difficult to find the encapsulated cells when they were lost among an excess of empty capsules.

Capsules were prepared also with a 50-50 poly(HEMA-MMA) copolymer dissolved in polyethylene glycol 600 (Figure 4). These capsules were approximately 0.9 mm in diameter with a 20 μm thick wall. However, polyethylene glycol has proven to be less than an ideal solvent: as the polyethylene glycol concentration (in the wash solution) was reduced to 10% (v/v), the encapsulated erythrocytes lysed regardless of the number of stepwise washes used. Since the encapsulating solution (60% (v/v) polyethylene gycol) was hypertonic it is presumed that the observed lysis was due to an osmotic pressure effect. Coextrusion of the two liquid phases (polyacrylate solution and freshly washed erythrocyte suspension) significantly reduced the lysis of poly(HEMA-MMA) microencapsulated red blood cells in an isotonic saline buffer. Presumably by reducing the contact time between the polyethylene glycol solution and the erythrocyte suspension, adverse cell-solvent interactions were minimized. Whether the concentrated polyethylene glycol solution sensitizes the cell membrane for subsequent lysis in more dilute solutions is currenty being investigated. The ability of polyethylene glycol to fuse cell membranes is well known[15] and may be relevant here.

Interfacial Adsorption

Calcium alginate immobilized erythrocytes were also encapsulated in a thin (~ 30 μm) EUDRAGIT film formed by interfacial adsorption. EUDRAGIT coated calcium alginate beads were large (~ 1.9 mm diameter) and contained normal, intact erythrocytes (Figure 5). In hematoxylin-eosin stained

46

histological sections of EUDRAGIT coated capsules, the capsule wall appeared as a densely stained band that is a consequence of polymer diffusion into the permeable, sponge-like gel network. A similar shell and core structure was apparent from scanning electron micrographs of freeze-fractured EUDRAGIT coated alginate capsules (Figure 6). Some alginate immobilized erythrocytes appeared as crenated discs, presumably due to the high extracellular calcium concentrations.

In other work, the presence of an adsorbed EUDRAGIT film on calcium alginate beads was demonstrated quantitatively by comparing the mechanical properties of uncoated and coated alginate beads in compression[16]. Calcium alginate capsules failed at a single value of the applied compressive load, the value of which was found to depend solely on the alginate content of the capsules, while EUDRAGIT coated calcium alginate capsules displayed two load maxima -- one characteristic of the alginate core, the other attributed to the presence of the EUDRAGIT coating.

The formation of stable aqueous emulsions from water-insoluble poyacrylates other than EUDRAGIT, the requirements for cell viability and growth in other hydrocolloid gel matrices (e.g., agarose and carrageenan) and the requirements for interfacial adsorption are currently under investigation.

Tissue Compatibility

Preliminary _in vivo_ biocompatibility studies of EUDRAGIT coated calcium alginate capsules were discouraging. Detailed histological sectioning of EUDRAGIT coated capsules after a 12 day intraperitoneal implantation in rats revealed a thick ($\sim$ 1 mm) fibrous capsule of well-developed scar tissue surrounded by polymorphonuclear granulocytes (Figure 7). The cationic nature of EUDRAGIT was implicated in the adverse tissue reaction[17] and it is expected that the neutral poly(HEMA-MMA) will be more tissue-compatible. On the other hand, the adverse tissue reaction may have been caused by the elution of contaminants (possibly aminated) from the implanted EUDRAGIT. A reprecipitated EUDRAGIT shows a lower toxicity towards fibrobasts in a cell culture assay system[6]. Extensive purification of EUDRAGIT is underway to reduce the severity of the tissue reaction.

Cell Viability

Results not presented here have demonstrated that functional erythrocytes were encapsulated in EUDRAGIT by either interfacial precipitation[5] or interfacial adsorption[6]. Glucose was consumed, albeit at a slightly higher rate than normal cells, and oxygen was reversibly bound to the cellular hemoglobin with P_{50} and n_{50} -- values comparable to those of stored blood after encapsulation. Although not conclusive evidence of cell viability, these are good indications that the encapsulated erythrocytes remain functional. Unfortunately, erythrocytes are largely senescent and the best assay of viability -- the clinical assessment of post-transfusion

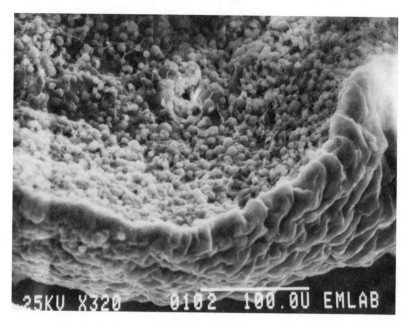

Fig.6. SEM of freeze-fractured EUDRAGIT RL coated calcium
        alginate immobilized erythrocytes.

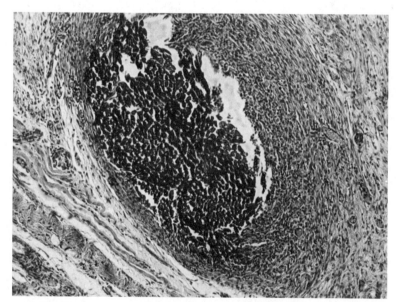

Fig.7. Hematoxylin-eosin stained histological section of
        sterilized EUDRAGIT coated calcium alginate capsules after
        12 days in the intraperitoneal cavity of rats.

viability — was impractical in this case. The more metabolically active, human fibroblasts are now being used as model cells to be better able to assessing the viability of encapsulated cells.

CONCLUSIONS

Microencapsulation of mammalian tissue cells by interfacial precipitation or interfacial adsorption of EUDRAGIT RL has led to the development of two microencapsulation strategies for the production of a durable implantable metabolic prostheses for the amelioration of insulin-dependent diabetes.

The production of microcapsules by interfacial precipitation of the water insoluble cationic polyacrylate, EUDRAGIT, has been refined. Capsules large enough to contain intact pancreatic islets have been prepared by a gentle, low shear coextrusion process. Soluble copolymers of hydroxyethyl methacrylate and methyl methacrylate have been synthesized and the preparation of more tissue compatible microcapsules with these polymers are currently under investigation.

The discovery that EUDRAGIT RL, although water insoluble, can be stable as an aqueous emulsion has lead to the development of an interfacial adsorption method for microencapsulation. EUDRAGIT coating of calcium alginate microcapsules, increased the compressive strength of the soft alginate, without adversely affecting the viability of the immobilized erythrocytes. Interfacial adsorption is currently being extended to other neutral water insoluble polycrylates where the process is expected to be governed not by polyelectrolyte complex formation but by the relative interfacial tensions of the capsule wall forming materials and the hydrocolloid gel.

The methods developed within the context of the development of a metabolic prostheses for diabetes are expected to be useful in other areas. New procedures for cell culture, cell product harvesting and drug delivery may be developed using these microencapsulation strategies.

ACKNOWLEDGEMENTS

The authors are grateful for the financial support of the Natural Sciences and Engineering Research Council and a joint grant of the Juvenile Diabetes Foundation with Dr. A.M. Sun of the Connaught Research Institute. The authors are grateful for the technical assistance of Frances Hincenburgs and the advice of Dr. A.M. Sun. One of us (MAW) acknowledges the receipt of a post-doctoral Fellowship from the Natural Sciences and Engineering Research Council.

REFERENCES

1. W.J. Spencer, IEEE Trans. Biomed. Eng. BME28:237 (1981).
2. Novo Industri A/S, "Infusion Pumps and Insulin Therapy", a technical brochure (1982).
3. M.V. Sefton, "Implantable Pumps", in: "Medical Applications of Controlled Release Technology", R.S. Langer and D. Wise, Eds., CRC Press (in press).
4. F. Lim and A.M. Sun, Science 210:908 (1980).
5. M.V. Sefton and R.L. Broughton, BBA 717:473 (1982).
6. F.V. Lamberti, M.A.Sc. Thesis, University of Toronto (1982).
7. G. O'Shea, H. Van Rooy, A. Wood and A.M. Sun, Diabetes 31(2): Abstract 203 (1982).
8. R.K. Gilding, personal communication.
9. G.T. Grant, E.R. Morris, D.A. Rees, P.C. Smith and D. Tham, FEBS Letts. 32:195 (1973).
10. G. Pilwat, P. Washausen, J. Klein and U. Zimmerman, Z. Naturforsch 35c:352 (1980).
11. "Tissue Culture; Methods and Applications", P. Kruse Jr. and M.K. Patterson, Eds., Academic Press (1973).
12. P.E. Lacy and M. Kostianovsky, Diabetes 16:35 (1967).
13. G.M. Healy, S. Teleki, A.V. Seefried, M.J. Walton and H.G. MacMorine, Appl. Microbiol. 21:1 (1971).
14. S.W. Thomson, in: "Selected Histochemical and Histopathological Methods", Charles C. Thomas, Pub., Springfield, Ill. (1966).
15. T.J. Aldvinkel, Q.F. Ahkwong, A.D. Bangham, D. Fisher and J.A. Lacy, BBA 689:548 (1982).
16. F.V. Lamberti and M.V. Sefton, BBA 759:81-91 (1983).
17. M. Barvic, J. Biomed. Mater. Res. 5:225 (1971).

MELT SPINNING OF POLY-L-LACTIDE AND

HYDROLYSIS OF THE FIBER IN VITRO

S.-H. Hyon, K. Jamshidi, and Y. Ikada

Research Center for Medical Polymers and Biomaterials
Kyoto University
Kawara-cho, Shogoin, Sakyo-ku, Kyoto 606, Japan

INTRODUCTION

It has been widely recognized that biocompatibility of a polymeric material is the minimal requisite for use as a surgical implant. What we mean by the biocompatibility is different from one implant to another, depending on the purpose and the duration of implantation. For instance, nonthrombogenicity is required for the material to be used as a small vascular graft, while strong bioadhesion with host tissues is needed if the polymer is used for artificial tracheas. On the other hand, soft biomaterials are often demanded which give no mechanical stimulus to the surrounding tissue, as IUD and contact lens. The biocompatibility involves all of these properties, implying that it is difficult to give a clear single definition to the term "biocompatibility".

However, one can say at least that such a material that invokes a considerable acute foreign-body reaction is not compatible with the biological environment. Acute inflammation, which is generally due to soluble molecules or ions of low molecular weight leached or degraded from the polymeric implant, can be mostly avoided if the polymer is properly chosen, constructed into the implant, and sterilized. It is, however, highly difficult to prevent all the foreign-body reactions, because the synthetic implant is more or less a material foreign to our body. To protect ourself, our body tries to wall the invading foreign material off or destroy it. Even if the response of our body to the foreign material is not obstinate in the initial stage of implantation, accumulation of the ever-continuing, physical and chemical stress by the implant might adversely affect the biological environment, resulting in significant injuring our living tissues after long-term implantation. In most cases the polymeric material undergoes encapsulation by a fibrous tissue when implanted as a tissue prosthesis. Lack of strong adhesion of the hard encapsulating tissue with the implant often invokes undesirable reaction to the living body like infection[1]. No change in size will be also a shortcoming of the implant when applied to children at the growing stage.

In this respect, biomaterials prepared from bioabsorbable polymers are of great interest, because the foreign implanted materials do not remain in body for ever, but disappear gradually or abruptly as a result of biodegradation[2]. However, at least two following conditions should be fulfilled so far as this material can be utilized as a surgical implant. One is to retain the minimal mechanical properties in body until the injured tissue will be completely cured. The other is not to produce any toxic compounds during biodegradation. One of the most promising candidates fulfilling these requirements seems to be a group of aliphatic polyesters, especially poly-L-lactide(P-L-LA), because this is crystallizable and L-lactic acid, the product of hydrolysis, enters into the metabolic pathways, resulting in excretion primarily as water and carbon dioxide[3]. It should be noted that polyglycolide and collagen have been clinically used as bioabsorbable polymers, but their rates of hydrolysis in body are somewhat too high to retain the strength for periods longer than half a month.

Therefore, we have selected P-L-LA as a starting polymer for surgical implants to be used as a temporary support in body up to curing. The purpose of spinning P-L-LA into fibers is that from the product we are able to construct many implants of various shapes and softness, such as thread, cord, gauze, knitted and woven clothes, unwoven fabrics, tube, and so on[4]. Although basic properties of P-L-LA like crystallization[5-8], microstructure[9,10], and crystalline structure[11,12] have recently been studied in detail, little is known on the hydrolysis[13] and biomedical applications of P-L-LA as fibers[12,14,15].

EXPERIMENTAL

P-L-LA

Synthesis of the L-lactide monomer and ring-opening polymerization of L-lactide were carried out following the method of Lowe[16] and Wasserman[17]. The starting L-lactic acid is a product of C.V. Chemie Combinatie Amsterdam C.C.A. (90 % aqueous solution, edible type) and was subjected to polycondensation without further purification. The oligomeric polycondensate was depolymerized in the presence of antimony trioxide (0.1 wt% of oligomer) to yield the lactide monomer. It was four times recrystallized from ethyl acetate. The purified monomer gave a melting temperature of 97.4 °C and a specific optical rotation of −270° in dioxane at 578 nm and 25 °C. The optical purity determined with lactic acid dehydrogenase was 97.5 %. Stannous octoate and lauryl alcohol were added to the monomer by 0.03 wt% and 0.01 wt% of the monomer, respectively, and the bulk polymerization was allowed to proceed at a reduced pressure of 10^{-3} mmHg. The resultant polymer was purified by precipitation into methanol from the chloroform solution. Hydroxyl endgroups were acetylated homogeneously at 60 °C for 5 hr in the 5 wt% chloroform solution containing 5 wt% acetic anhydride and 2.5 wt% pyridine. The viscosity-average molecular weight of polymers was determined from the intrinsic viscosity in chloroform at 25 °C by using the equation[18],

$$[\eta] = 5.45 \times 10^{-4} \times \overline{Mv}^{0.73}$$

Melt spinning and hot drawing

The acetylated polymer was melt-spun in air at 200 °C with an extruder of screw-type. The diameter of the screw was 25mm and the nozzle had ten holes of 1.0 mm diameter. The spun fibers were drawn to various ratios in air at temperatures from 80 to 160 °C.

Heat treatment of the drawn fibers was done in the dry air for 10 min at temperatures from 80 to 160 °C.

Measurement of physical properties of fiber

X-ray diffraction patterns were photographed with a wide angle X-ray flat camera and a small angle X-ray camera manufactured by Rigaku Denki Co., using Cu-kα (Ni filtered) at 50 kV and 80 mA. Birefringence was measured with a polarization microscope provided with a Berek compensator. The fiber density was determined at 30 °C in a density-gradient tube consisting of n-hexane and carbon tetrachloride. The temperature dependence of the dynamic modulus E' and the dynamic loss E" was measured with a Vibron DDV-2 of Toyo-Baldwin Co., a type of forced vibrational method. The frequency selected for the measurement was 110 cps throughout this work and the values E' and E" were determined in a range of temperature from 20 to 160 °C at a heating rate of $1 °C \cdot min^{-1}$. Stress-strain curves were recorded for fibers of 20 mm length at an elongation rate of $20 mm \cdot min^{-1}$ with Tensilon UTM-II manufactured by Toyo-Baldwin Co. The temperature and relative humidity at the measurement were 20 °C and 65 % RH, respectively.

Hydrolysis of P-L-LA fibers

The *in vitro* hydrolysis was carried out at 37 and 100 °C in phosphate buffer saline of pH 7.4 for the fibers six times drawn at 80 °C, followed by heat treatment for 10 min at 140 °C. For comparison, Dexon-S® was also subjected to hydrolytic degradation under the same condition. The fibers hydrolyzed were observed with a scanning electron microscope MSM-9 manufactured by Hidachi-Akashi Co. after coating with platinum.

RESULTS AND DISCUSSION

Properties of P-L-LA fibers

The polymerization condition of L-lactide and characteristics of the polymer used for the melt spinning are summarized in Table 1. Figure 1 shows X-ray diffraction patterns of the fiber obtained by spinning at 200 °C at a draft ratio of about 25. Photos (A), (B), and (C) in Fig.1 are ,respectively, the

Table 1. Polymerization Conditions and Characteristics
of the Polymer used for Melt Spinning

Polymerization	
$[\alpha]_{578}^{25}$ (dioxane) of L-lactide	-270°
T_m of L-lactide	97.4 °C
[stannous octoate][a]	0.03 wt% of monomer
Polymerization temperature	140 °C
Polymerization duration	12 hr
Monomer conversion	97 %
Resulting polymer	
$[\alpha]_{578}^{25}$ (dioxane)	-170°
$\overline{M}v$	3.6 x 10⁵
T_g (DSC)	57 °C
T_m (DSC)	184 °C

a Lauryl alcohol was added to the catalyst by 0.01 wt% of monomer.

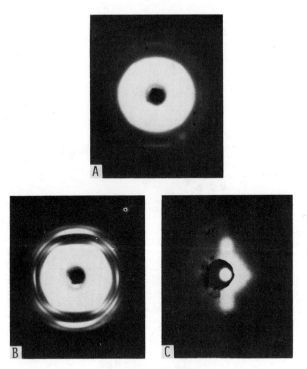

Figure 1. X-ray patterns of the undrawn fibers: (A) wide angle
X-ray diffraction pattern of the fiber as spun; (B) wide
angle X-ray diffraction pattern of the fiber heat-treated at 140 °C for 10 min in the free state; (C) small angle
X-ray scattering pattern of the fiber heat-treated as (B)

wide angle diffraction pattern of the fiber as spun, the wide
angle diffraction pattern of the fiber heat-treated at 140 °C
for 10 min in the free state, and the small angle scattering
pattern of the fiber heat-treated as above. As is seen, the
pattern in Fig.1A is hallow as a whole, characteristic of amorphous
polymers but also has peaks due to the oriented crystalline
region, though weak. The pattern exhibiting more strongly
oriented crystalline region is observed in Fig.1B. The small
angle X-ray scattering result in Fig.1C has a discrete two-
point pattern to imply the oriented lamellae structure. Melt
spinning of the P-L-LA at 200 °C resulted in reduction in $\overline{M}v$
from 3.6×10^5 to 1.1×10^5.

Figure 2 shows the result of dynamic viscoelastic measurement on the fiber undrawn but heat-treated at 140 °C for 3 hr.
It is seen that the dynamic modulus E' has such a high value
as 8.3×10^{10} dyne·cm^{-2} at room temperature but exhibits a steep
decrease occurring from about 50 °C. On the other hand, a maximum

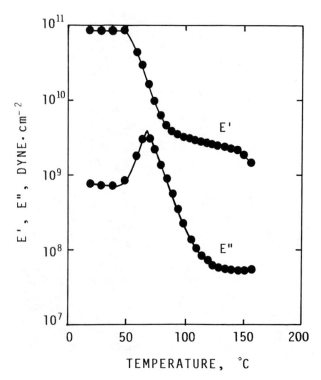

Figure 2. Dynamic modulus E' and loss E" plotted against temperature for the undrawn fiber heat-treated at 140℃ for 3 hr

is observed at 66.5 °C for the dynamic loss E". This absorption may be ascribed to micro-Brownian motion of the main chain of P-L-LA and hence corresponds to the glass transition. Therefore, the fiber drawing was always conducted at temperatures higher than this glass transition point. Change in birefringence by drawing at 80, 140, and 160 °C is shown in Figure 3. In every case, birefringence is increased with the draw ratio, but tends to level off from a draw ratio around 6.

Figure 4 shows the wide angle X-ray diffraction pattern of the fiber drawn at 80 °C, followed by heat treatment at 140℃ in the fixed state. Clearly, the crystalline orientation is observed even for the undrawn fiber (the draw ratio is unity), but becomes more dominant as the draw ratio is increased. The wide angle X-ray fiber diagram of P-L-LA has been also reported by other workers [11,12]. According to Pennings and his coworkers [12], the oriented P-L-LA has two crystalline forms; one is α form with helix conformation and the other is β form with an extended helix conformation. The diffraction pattern of our fiber suggests it to be the α form, based on their findings.

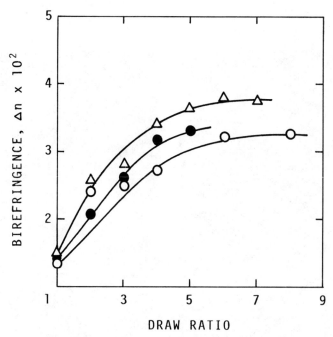

Figure 3. Dependence of birefringence on draw ratio for the fiber drawn at different temperatures: (O) 80 °C; (●) 140 °C ; (Δ) 160 °C

Densities of the fiber drawn at different temperatures are plotted as a function of the draw ratio in Figure 5. We could not determine accurately the density of the fiber drawn around 80 °C because of void formation during drawing. However, such voids were not formed on the fibers when drawn at 140 and 160 °C even at high draw ratios. In Fig.5 is also shown the crystallinity of drawn fibers estimated from the observed density using 1.290 and 1.248 $g \cdot cm^{-3}$ as densities of the crystalline and amorphous regions[5], respectively. Detailed discussion on the crystallinity of the fiber needs accumulation of much more experimental results.

Figure 6 shows tensile strengths of the fiber drawn to different elongations, followed by annealing. The strengths for the knotted fiber are also given in Fig.6. It can be seen that fibers of higher tensile strengths are obtained as the draw ratio becomes higher. Rise in temperature at drawing gives the fiber with higher strength, in correspondence with higher density and birefringence. In the present case, the optimum temperature for drawing was 160 °C and drawing was impossible at temperatures higher than 160 °C. This reason may be due to partial melting of the crystalline region which begins

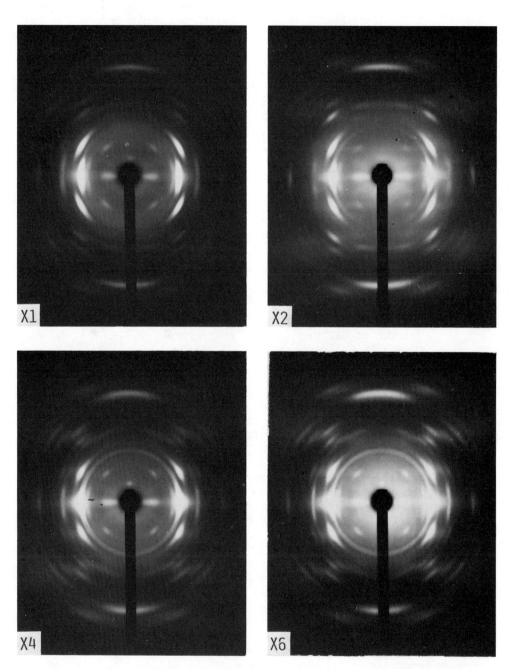

Figure 4. Wide angle diffraction patterns of the fiber drawn
at 80 °C, followed by annealing at 140 °C. The draw
ratios are given in the photos

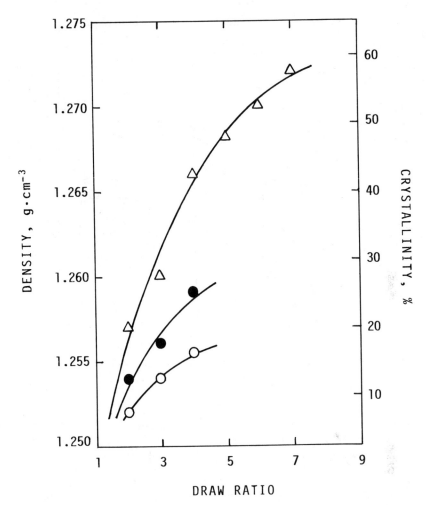

Figure 5. Densities and crystallinities of the fiber drawn at different temperatures; (O) 120 °C; (●) 140 °C; (Δ) 160 °C

at 160 °C, as a differential scanning calorimetric study indicated (the result is not shown here). Figure 7 shows the Young's moduli and elongations at break evaluated from stress–strain curves of the fibers. The tensile strengths and moduli of these fibers are comparable to those of polyglycolide fiber[19].

Hydrolysis *in vitro*

No appreciable change both in tensile strength and weight loss was detected when the fiber six times drawn at 80 °C followed by heat treatment for 10 min at 140 °C was immersed in phosphate buffer saline at 37 °C for 6 months. Therefore, the temperature

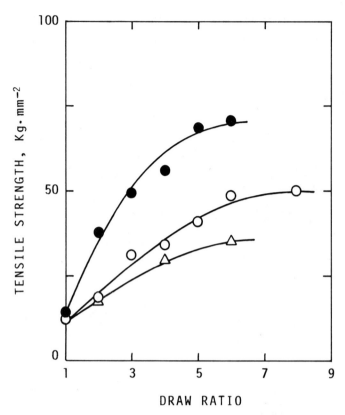

Figure 6. Tensile strength of the fiber drawn at different Temperatures, followed by annealing at 140 °C; (O) 80°C; (●) 160 °C and (Δ) 80 °C for the knotted fiber

of buffered solution was raised to 100 °C to accelerate the hydrolysis. The result is given in Figure 8, together with that for Dexon®. As is apparent, Dexon® shows reduction in tensile strength by 50 % after 1 hr and in weight loss by 100 % after 4 hr, while the same reduction is observed after 10 and 40 hr, respectively, for the P-L-LA fiber. Simply, under this accelerated condition, the P-L-LA fiber seems to have a hydrolysis rate ten times as low as that of Dexon® .

SEM examination for the fiber subjected to hydrolysis at 37 °C for 6 months exhibited no change, in correspondence with no physical change of the fiber. However, the fiber hydrolyzed at 100 °C demonstrated a significant change in the SEM.

In Figure 9 is shown the result of SEM study for Dexon® (A, B, C in the left panel) and for our P-L-LA fiber (D, E, F in the right panel). Both were hydrolyzed in buffered solution

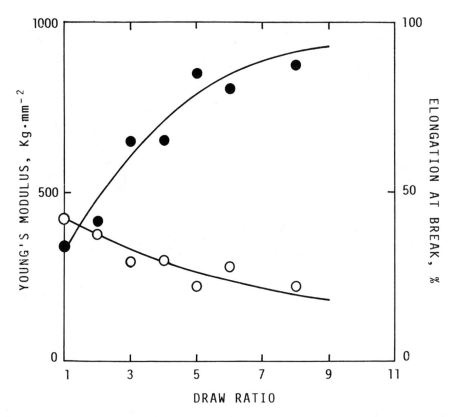

Figure 7. Young's modulus and elongation at break for the fiber
drawn at 80 °C, followed by annealing at 140 °C

kept at 100 °C and the durations were 0 hr (A), 1 hr (B), and 2 hr
(C), 0 hr (D), 10 hr (E), and 20 hr (F). As is obvious, no significant
difference in appearance is observed between the upper and
the middle pictures for both fibers, though the tensile strengths
are reduced to half of those of intact fibers. However, after
hydrolysis for 2 hr, small cracks are seen all over the surface
of Dexon® in the vertical direction along the fiber axis.
Such a regular pattern of surface cracks has also been observed
by Chu and Campbell[20] on the Dexon® which was irradiated with
γ-rays to 40 Mrad, followed by hydrolysis *in vitro* at 37 °C
for 40 hr.   Interestingly, they did not observe such regular
cracks for the unirradiated Dexon® fiber even when it was subjected
to hydrolysis for 90 day at 37 °C.   Regular cracks appear for
the P-L-LA fiber subjected to hydrolytic degradation at 100 °C
for 20 hr as is seen in Fig.9 F.

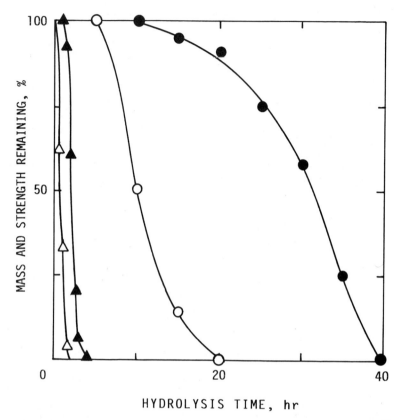

Figure 8. % Decrease in tensile strength and mass for Dexon® and
the P-L-LA fiber hydrolyzed at 100°C; (Δ) strength of
Dexon®; (▲) mass of Dexon®; (O) strength of P-L-LA;
(●) mass of P-L-LA

In conclusion, we may summarize that melt spinning of
the acetylated P-L-LA can produce the fibers having tensile
strengths comparable to conventional crystalline polymers and
much high resistance to hydrolytic degradation in comparison
with the fiber from polyglycolide. It is expected that bioabsorbable fibers with various hydrolytic rates will be obtained
by modification of P-L-LA, for instance, with copolymerization,
blending, or coating. The biological responses to implants
prepared from the P-L-LA fiber will be described in the near
future.

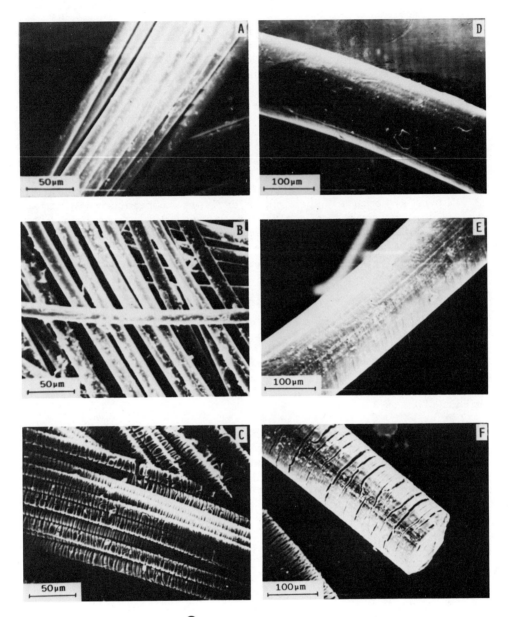

Figure 9.   SEM of Dexon®(A,B,C) and the P-L-LA fiber (D,E,F)
            hydrolyzed at 100°C; (A) 0 hr; (B) 1 hr; (C) 2 hr; (D)
            0 hr; (E) 10 hr; (F) 20 hr

ABSTRACT

    L-Lactide was polymerized in bulk at 140 °C to yield poly-
L-lactide with molecular weight of 3.6×10^5.   Melt spinning
of poly-L-lactide was conducted in air at 200 °C after acetylation
of the terminal OH groups of the polymer.   The tensile strength
of the fibers, drawn and annealed at 160 °C, was 72 kg·mm^{-2} for
a draw ratio of 6.   The X-ray diffraction pattern was diffuse
for the undrawn fiber, whereas it became very sharp upon drawing
and annealing, indicating that typical, highly oriented fiber
could be obtained from the poly-L-lactide.   The crystallinity
of fibers obtained was estimated from the fiber density to
be in the range from 5 to 55 %.   The poly-L-lactide fiber practically underwent no degradation in phosphate buffer saline at
37 °C over 6 months.   When the hydrolysis temperature was raised
to 100 °C, the weight of poly-L-LA was reduced to the half after
30 hr, while Dexon® , a commercially available polyglycolide
suture, was degraded to the same extent after 2 hr when it
was subjected to the hydrolysis condition identical to that
of poly-L-lactide.

Acknowledgement

    We gratefully acknowledge the help of Drs. T. Kitao and
Y. Kimura of Kyoto Institute of Technology in fiber spinning.

REFERENCES

1. L.L. Hench, *Science*, 208, 826 (1980).
2. R.L. Kronenthal in *"Polymers in Medicine and Surgery"*,
    R.L. Kronenthal, Z. Oser, and E. Martin, Eds., Plenum Press
    New York, 1975, p.119.
3. M. Vert, F. Chabot, J. Leray, and P. Christel, *Makromol.
    Chem.*, *Suppl.* 5, 30 (1981).
4. A.S. Hoffman, *J. Appl. Polym. Sci. Appl. Polym. Symp.*,
    31, 313 (1977).
5. E.W. Fischer, H.J. Sterzel, and G. Wegner, *Kolloid-Z.u.Z
    Polymere*, 251, 980 (1973).
6. B. Kalb and A.J. Pennings, *Polymer*, 21, 607 (1980).
7. R. Vasanthakumari and A.J. Pennings, *Polymer*, 24, 175 (1983).
8. F. Chabot, M. Vert, S. Chapelle, and P. Granger, *Polymer*,
    24, 53 (1983).
9. C.V. Prasad and K. Sundram, *Int. J. Quantum Chem.*, 20,
    613 (1981).
10. R.J.M. Zwiers, S. Gogolewski, and A.J. Pennings, *Polymer*,
    24, 167 (1983).
11. P.D. Santis and A.J. Kovacs, *Biopolymers*, 6, 299 (1968).

12. B. Eling, S. Gogolewski, and A.J. Pennings, *Polymer*, 23, 1587 (1982).
13. L. Sedel, F. Chabot, P. Christel, X.d. Charentenay, J. Leray, and M. Vert, *Rev. Chir. Orthop.*, 64 Suppl. II, 92 (1978).
14. Y.M. Trehu (Ethicon, Inc.), *U.S.P.* 3,531,561 (1970).
15. A.K. Schneider (Ethicon, Inc.), *U.S.P.* 3,636,956 (1972).
16. C.E. Lowe (E.I. du Pont de Nemours & Co., Inc.), *U.S.P.* 2,668,162 (1954).
17. O. Wasserman and C.C. Versfelt (Ethicon, Inc.), *U.S.P.* 3,839,297 (1974).
18. A. Schindler and D. Harper, *J. Polym. Sci., Polym. Chem. Ed.*, 17, 2593 (1979).
19. C.C. Chu, *J. Appl. Polym. Sci.*, 26, 1727 (1981).
20. C.C. Chu and N.D. Campbell, *J. Biomed. Mater. Res.*, 16, 417 (1982).

SOME MORPHOLOGICAL INVESTIGATIONS ON AN ABSORBABLE COPOLYESTER

BIOMATERIAL BASED ON GLYCOLIC AND LACTIC ACID

Brandt K. Carter and Garth L. Wilkes

Department of Chemical Engineering and
Polymer Materials and Interfaces Laboratory
Virginia Polytechnic Institute and State University
Blacksburg, VA 24061-6496

ABSTRACT

Some morphological studies were carried out on a physiologically absorbable copolyester material. While this material has been utilized as a suture material under the trade name of Vicryl[R] and possesses the potential for utilization in other biomaterial devices such as controlled drug release matrices, little information is available on its morphological state. Specifically the details of the crystalline morphological textures and their variation with thermal history was the primary topic of interest for this study. A chemical etching methodology was developed for purposes of enhancing structural details of this semicrystalline material. By using a refined hydrolysis procedure, the amorphous phase can be preferentially degraded and removed thereby enhancing the details of the underlying crystalline morphology. The particular work presented here focuses on studies that have utilized specially prepared spherulitic films of this copolymer. The results emphasize structural details of partially spherulitic films as well as totally spherulitic films possessing different thermal histories. In addition, these same chemical etching procedures were used to investigate the uniaxial and biaxially deformation of these spherulitic superstructures. It is suggested that similar methodology might well be utilized for investigating similar hydrolizable polyester biomaterials for purposes of structural investigations.

INTRODUCTION

Over the last two decades there has been a considerable growth of interest in the development and subsequent application of absorbable biomaterials. Two principal examples are the absorbable materials provided by Ethicon Incorporated (glycolic/lactic acid copolymers) and American Cynamide (polyglycolic acid). The application of these two specific materials as sutures and potentially controlled drug release matrices is made possible by the fact that these polymers slowly hydrolize in the presence of the physiological environment and that the resulting by-products can be metabolized. To date there has been little structural work carried out on these materials in published form. Thus, the authors have carried out some initial structural investigations on the absorbable copolyester materials based on the glycolic and lactic acid material as supplied by Ethicon. It is not the aim of this paper to direcly correlate the details of structural analysis, as noted by electron microscopy, to the nature of the physiological hydrolysis rates, potential drug transport considerations or the affect on mechanical properties. Rather, it is simply to present some of the preliminary investigations that we have carried out regarding this particular system as a useful model polymer for studying semicrystalline textures produced in quiescently crystallized materials. However, the relevance of many of the considerations given within this paper clearly have impact on such features as potential drug release behavior, general mechanical behavior of the resulting materials and undoubtedly other aspects which relate to the utilization of similar semicrystalline polymers as potential biomaterial candidates.

As just stated, this polymer is somewhat viewed as a model system for these investigations since it is particularly suceptable to hydrolytic attack by design. This allowed the preferential chemical attack of the amorphous phase in a relatively simple manner and degraded this phase faster than the crystalline phase. This promoted contrast between the crystalline and amorphous phases necessary for the electron microscopy investigations. Certainly chemical etching procedures of the hydrolytic type have been applied to polyester materials in the past but we feel that the studies presented will definitely illustrate that the particular polymer chosen for the work sited here offers considerable advantage in promoting this contrast between crystal and amorphous textures.

EXPERIMENTAL

Materials: The copolyester of glycolic and lactic acid, was supplied through the courtesey of Ethicon Inc. The material as received was in pellet form. The monomer ratio was approximately 10% lactide and 90% glycolide. This particular polymer and its general characteristics has been discussed by Kronenthal et al.[1]. Typically it is utilized as a semicrystalline material that possesses a melting point of about 200°C and a glass transition temperature of about 43°C. With respect to the general characteristics of this polymer, it is a slow crystallizing macromolecule when can be easily quenched directly into the glassy state by rapid cooling from the melt.

Film Preparation: A small laboratory extruder (Maxwell Melt Elastic Extruder) was used to homogenize the as received polymer after first being sure that the material had been kept free from moisture uptake. The extrudate at 220°C was directly placed upon aluminum foil sheets between which the copolyester was then compression molded into films at 260°C immediately after extrusion. The films were isothermally crystallized from the melt at specific temperatures in the range between the melting point and the glass transition temperature. After crystallization had reached the desired level the film which was still between foil sheets, was quenched in ice water. This procedure generated isolated spherulites in an amorphous matrix with a surprisingly monodisperse distribution of spherulite diameters. The quenching into icewater caused no promotion of hydrolysis since the material was still between the aluminum foil sheets at that time. Following removal from the water bath, the material was stripped from the aluminum foil sheets. Alternatively, the amorphous matrix surrounding the large isolated spherulites could be converted to a matrix of volume filling smaller spherulites (10-20 m) by simply raising the films from below Tg (room temperature) to 100°C for 10 minutes.

Deformation: Uniaxial deformation of isolated spherulites in predominantly amorphous films was carried out at ambient temperatures at various elongational rates. In order to effect the biaxial deformation of the large isolated spherulites, the surrounding amorphous matrix was converted to the volume filling smaller spherulites. This served to reinforce the matrix and more closely match the mechanical characteristics of large spherulite and matrix. Balanced biaxial deformation of this fully spherulitic material was carried out above Tg at various elongational rates. An Instron Model 1122 was used for the uniaxial studies while a T.M. Long stretcher was used for the biaxial deformation.

Etching Technique: The use of chemical etching has long been used in the area of polymer morphology for purposes of preferentially degrading one specific component in a multicomponent system. Following this selective degradation, subsequent investigation and characterization of the remaining structure is typically made through the use of microscopy. Two general reviews that have discussed several of the etching procedures utilized in morphological studies of bulk crystalline polymers have been given by Wunderlich[2] and Hobbs[3]. By far and away the largest body of work dealing with chemical etchant enhancement of semicrystalline morphology has been the use of mineral acids on polyolefins[4]. One of the first chemical etching techniques which was developed for the analysis of cellulose was based on hydrolysis chemistry[5]. Since that time exhaustive acid and base hydrolysis of polyesters and polyamides have been reported[6]. Of particular interest to these authors is the study of Fischer et al.[7] on the base hydrolysis of polylactides wherein a 1:2 water-methanol mixture containing 0.02 to 0.04 moles per liter of sodium hydroxide was used to cleave the ester linkages at 20°C.

While the various chemical reagents utilized in the above studies are somewhat different depending on the nature of the polymer, the overall objective is to chemically attack the amorphous or less ordered phase more rapidly than that of the crystalline phase. A significant difference in these reaction rates will lead to surface texture contrast in bulk crystallized samples that can be easily observed with a scanning electon microscope. It is this same goal that we have tried to develop through the use of a suitable hydrolysis reaction for the copolyester described. As indicated earlier, the choice of this polyester was in fact based on the recognition that it does undergo pronounced hydrolysis.

The standard etching procedure that was utilized is as follows: a given sample was presoaked in dimethylsulfoxide (DMSO) at room temperature for between one and two minutes. This served to swell the remaining amorphous material and make it more suceptable to subsequent etching. Upon removal from the DMSO the sample was immediately immersed in a 50/50 solution of DMSO and distilled water which had been brought to a saturation condition with sodium carbonate at 70°C thereby making the etchant fluid alkaline. The duration of the etch varied from 2 to 10 minutes depending on the various thermal and deformation histories of the sample. No specific recipe can be quoted here since it was found that some variation in procedures was occassionally necessary as the nature of the crystalline morphology varied from sample to sample. More detailed structures could often be revealed through a modified procedure whereby the etching was performed at room temperature without the afore mentioned presoak. In this case the duration of the etch was varied between 10 hours and three weeks and was accompanied by a slow agitation. Following etching, each sample was then washed in distilled water at room temperature for 5 minutes and dried under vacuum for at least 24 hours.

Scanning Electron Microscopy: Following drying and metallization by sputter coating, the morphological texture of the etched materials were observed with an ISI Super 3-A scanning electron microscope.

RESULTS AND DISCUSSION

For purposes of illustrating the utility of the chemical etching reagents developed through this study Figure 1 is shown for later purposes of contrast. It is a SEM photo of a typical film surface containing isolated spherulites that are embedded within an amorphous matrix as developed through the isothermal crystallization procedure. As can be observed, a disc-like spherulite resides within the surrounding amorphous material. Without any chemical treatment there is no discernable superstructure within a spherulite. Even under increased magnification of Figure 1 the only texture that is noticable is

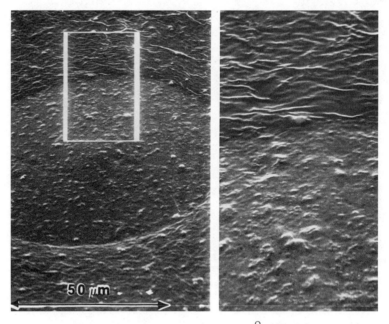

FIGURE 1. Spherulite crystallized at 160°C, unetched film surface. Magnification 1000x, insert magnification 3000x.

that left by the impression of the foil surface in the molded film.

Figure 2A and B present two additional scanning electron micrographs which show a spherulite embedded in an amorphous matrix which has been given the standard etching treatment. The discrete radial texture of the spherulite is now well accentuated although not at the lamalla level. Since the film thickness (125 micrometers) is less than the overall observed spherulite diameter (∿200 micrometers) it is clear why the spherulites appear more as disk like. In this particular case the spherulite is nucleated somewhat below the surface of the film and has undergone classical spherulitic growth. With increasing diameter the spherulite growth has terminated at the film surface and continued growth was essentially two dimensional in the plane of the film. The growth mechanism was apparent from the radial texture of the spherulite which terminates normal to the film surface near the spherulite center as shown in Figure 2A while Figure 2B displays more of the fibril textures being somewhat

FIGURE 2a. Spherulite crystallized at 160°C after the standard etch treatment. Split image highlights the fibrillar structure normal to the film surface at the spherulite center. Magnification 300x, insert magnification 3000x.

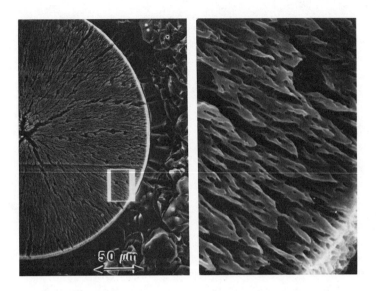

FIGURE 2b. Spherulite crystallized at 160°C after the standard etch treatment. Split image highlights the fibrillar structure parallel to the film surface at spherulite radial extremity. Magnification 300x, insert magnification 3000x.

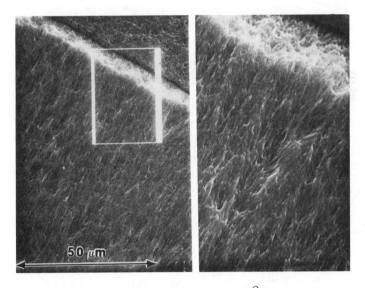

FIGURE 2c. Spherulite crystallized at 160°C after the modified low temperature etch treatment. Split image highlights the fibrillar structure at the spherulite radial extremity. Micrograph shows boundary between spherulite and surrounding amorphous matrix. Magnification 1000x, insert magnification 3000x.

more parallel to the film surface at longer radii as would be expected. It is also clear from these same two figures that the crystallization has led to a very uniform spherulite boundary between the edge of this superstructure and initially amorphous matrix. It should be pointed out here that the initially amorphous matrix now has a honeycomb like structure which we are strongly certain results from the initial solvent induced crystallization caused by the etching liquids and the subsequent etching of this specific area of the material. Further details with respect to the honeycomb structure of this solvent induced crystallized area will not be discussed further in this paper although we hope to discuss this subject at a later time in subsequent publications. In contrast, Figure 2C shows a similar spherulite which has undergone the modified etching procedure. Clearly, the severity of the etch is greatly reduced and consequently the level of structure discernible is of a finer nature.

One of our goals was to obtain some information regarding the overall interaction of spherulites with one another, i.e. what is the nature of the crystalline fibrillar interaction at interspherulitic boundaries? Figures 3A, A and B we note the

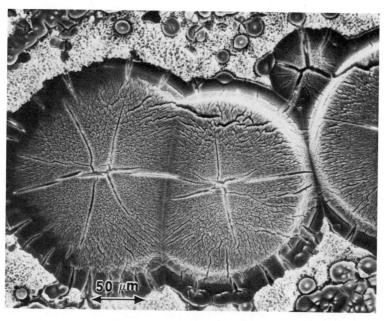

50 μm

FIGURE 3a. Two impinging spherulites crystallized at 120°C after standard etch treatment. Magnification 500x.

74

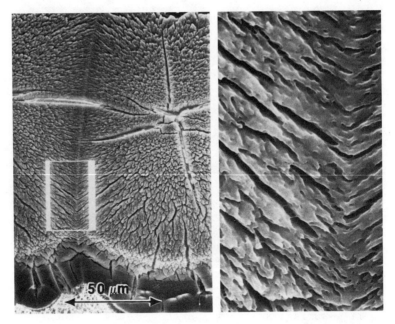

FIGURE 3b. High magnifications of the two impinging spherulites
showing the spherulite interface region. Magnification 700x,
insert magnification 3500x.

very well defined interface between these spherulites as given
within the higher magnified region. It is clear from these
micrographs that there appears to be good interfacing of the
crystalline fibrils of one spherulite with those of neighboring
spherulites. While we are not attempting to indicate that this
"interface" occurs at the lamallae level, the micrographs do
suggest that there is a general connectivity of the fibrils. In
particular, the reader notes that upon etching, the
interfibriller material is chemically removed and the remaining
troughs or voids often extend from between two radial fibrils of
one spherulite to between two fibrils of a neighboring
spherulite. This is particularly obvious when the fibrils of the
two spherulites are lying within the same plane and opposing each
other. Even in the case where these fibrils are adjoining
another spherulite in a more orthogonal approach (see Figure
4A-C) there still appears to be considerable continuity across
the interspherulitic boundaries possibly suggesting that good
interconnectivity of the fibrils between spherulites can also
occur even when they approach each other in a non aligned manner.

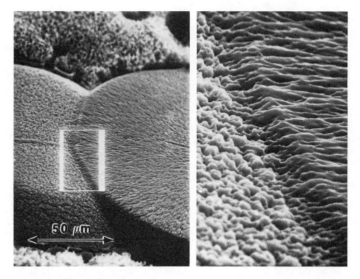

FIGURE 4a. Three impinging spherulites crystallized at 140°C after the standard etch treatment. Two spherulites were nucleated well below the film surface and show fibrils normal to the film surface. The third spherulite was nucleated at or very close to the film surface and show fibrils parallel to the film surface. 45° stage tilt. Magnification 500x, insert magnification 2500x.

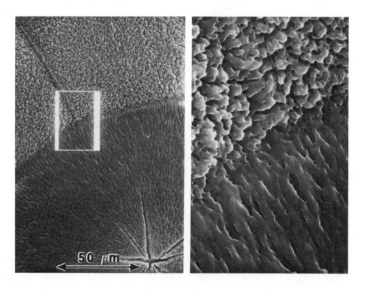

FIGURE 4b. The same trio of impinging spherulites shown in Figure 4A except the stage tilt is set at 0°. Magnification 500x, insert magnification 2500x.

76

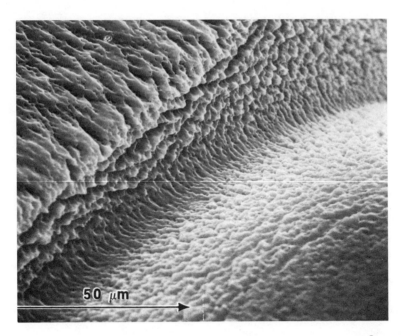

FIGURE 4c. Two impinging spherulites crystallized at 160°C after standard etch treatment. Clearly the fibrils exhibit a continuity across the spherulite interface dispite the orthogonallity of the fibrils from the adjacent spherulites. Magnification 2000x.

It might be mentioned that the authors do not wish to convey the thought that such good connectiveness occurs in the cases of all interspherulitic boundaries of all spherulitic materials. Specifically, there have been studies which indicate that when the spherulitic crystallization develops from a system containing impurities such as atactic species, or considerable low molecular weight fractions, this noncrystallizable material will often be concentrated at the interspherulitic boundaries and may lead to poorer fibrillar interaction across the boundaries[8,9]. Thus, we simply wish to point out for the system we are discussing here, there appears to be considerable interfacing of the crystalline fibrils from one spherulite to another can occur in the boundary regions and seems independent of the angles at which they may approach each other. Further detailed studies utilizing STEM will hopefully provide further indepth characterization of the boundary regions. Figures 5A, B & C shows an area of multi-spherulitic impingement. All three levels of magnification

FIGURE 5a. Area of multi-spherulite impingement after standard etch
treatment. Magnification 300x.

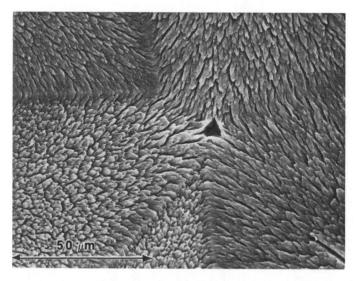

FIGURE 5b. Higher magnification of multi-spherulite impingement
shown in Figure 5a. Magnification 2000x.

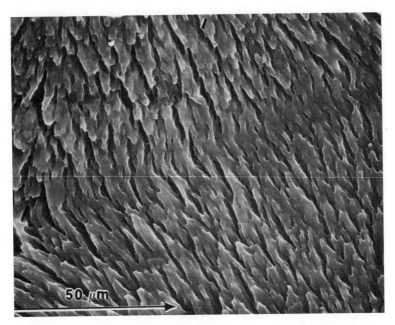

FIGURE 5c. Higher magnification of multi-spherulite impingement shown in Figure 5a. Magnification 2000x.

give a clear indicatioin of good interspherulitic interfacing through the radial fibrils. It is particularly interesting to note that at the very vertex of where several of these spherulite boundaries come together, a "void" is distinctly noted of a size that is large relative to that of the inter fibrillar regions where amorphous material had resided. We suspect that this void region comes as a result of not only an amorphous fraction that had resided at this point but also likely from the fact that there may have been actually partial void area in this region due to the densification that occurred during the initial crystallization. Such void formation has been noted by others in crystallization studies of polypropylene (8). A similar multi-spherulitic impingement area is shown in Figure 6. In this case however, the modified low temperature etching technique was used to highlight the crystalline textures in contrast to the results of the standard etching just described. It is quite clear that as one observes these specific micrographs, that there is little doubt about the overall crystalline fibril interaction.

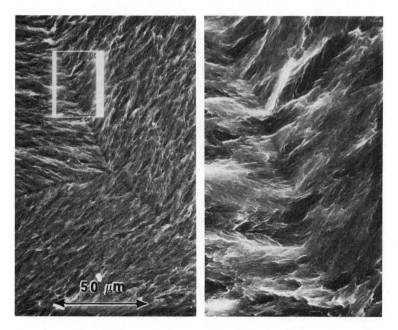

FIGURE 6. Area of multi-spherulite impingement after modified low
temperature etch treatment. Magnification 500x, insert
magnification 2500x.

Crystallization temperature has been shown to influence the
nature of spherulitic morphology[10].   Figures 7A,B & C documents
this effect in this copolyester system. Each micrograph depicts
a   spherulite   isothermally grown at a different level of
supercooling. The crystallization temperatures chosen range from
120°C to 190°C and were within the nucleation controlled region
of the growth rate-temperature spectrum for this system. It has
been well indicated through earlier studies such as those by
Keith and Padden[11] that changing crystallization temperature
can strongly alter the coarseness of the fibrillar development
within spherulites. We can clearly see this change in coarseness
as the crystallization temperature is increased. It is noted
that again interspherulitic boundaries display very   strong
interfacing of the crystalline fibrils at least at the level of
magnification provided here. It should be commented, however,
that at the particularly high crystallization temperatures, when
one investigates the fibril structure at a higher magnification,
some particularly fine threadlike interconnective bundle like
regions exist again indicating changes in the texture at   a   much

FIGURE 7a. Spherulite crystallized at 190°C after standard etch
treatment. Magnification 500x, insert magnification 2500x.

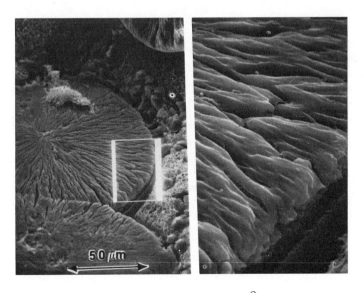

FIGURE 7b. Spherulite crystallized at 180°C after standard etch
treatment. Stage tile 45°. Magnification 500x, insert
magnification 2500x.

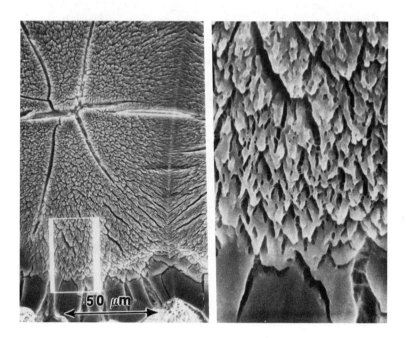

FIGURE 7c. Spherulite crystallized at 120°C after standard etch treatment. Magnification 500x, insert magnification 2500x.

finer scale. A similar fine texture can be easily seen in Figures 8A,B &C where spherulites of the same isotherms have been etched by the modified low temperature technique.

From the point of view of the biomaterials applications of this material or related semicrystalline materials that have undergone quiesent crystallization, it should be obvious that the parameter of crystallization temperature or more specifically the degree of supercooling is a very important consideration since the overall morphological texture is strongly influenced accordingly. With respect to such considerations as drug transport through such semicrystalline matrices or, in fact, even the rate of hydrolysis of such a matrix, this same fabrication variable will undoubtedly play an important role although we have not systematically investigated this behavior.

Turning briefly to the effect of deformation on spherulitic textures and how it may be of some relevance in the area of biomaterials, Figure 9 illustrates at low magnification a spherulite which has undergone uniaxial extension at the

82

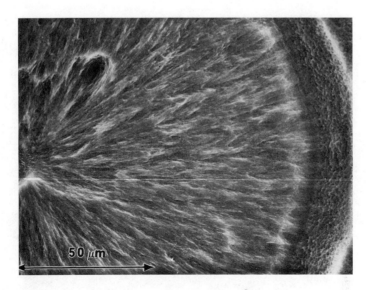

FIGURE 8a. Spherulite crystallized at 190°C after the modified low
temperature etch treatment. Magnification 3000x.

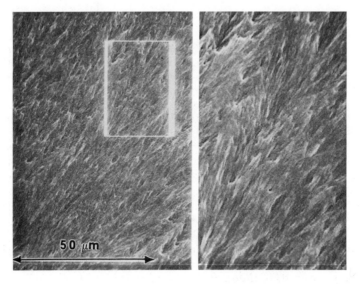

FIGURE 8b. Spherulite crystallized at 180°C after the modified low
temperature etch treatment. Magnification 1000x, insert
magnification 3000x.

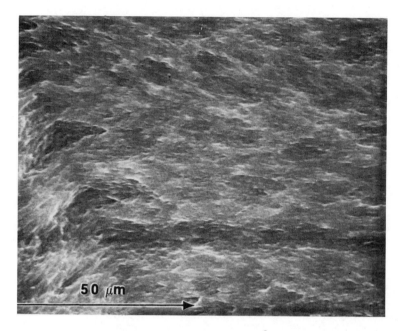

FIGURE 8c. Spherulite crystallized at 120°C after the modified low
temperature etch treatment. Magnification 3000x.

conditions indicated within the figure caption. In this case the
large spherulite is immersed within a matrix of amorphous
material as has been described in the experimental part.   It is
important   to point out that in these systems, the large
spherulite that can be denoted has undergone rather typical
uniaxial deformation. Specifically, it has been transformed from
a circular shape to that of an ellipsoidal shape. The goal here
is again not to deleniate indepth details of   spherulitic
deformation but to point out some of the general behavior that
occurs in these systems and its possible relevance to biomaterial
considerations - the latter point being addressed below.   The
micrographs of Figures 10A & B show the structural details of an
uniaxially deformed spherulite as enhanced by our standard
etching techniques.   These observations are very much in line
with the overall sketch shown in Figure 11 which is a simplified
model of spherulitic deformation. Various regions of a
spherulite deform differently as dictated by the direction of the
applied   stress   relative   to   the fibril axes and related
interstitial amorphous material. This model is not particularly

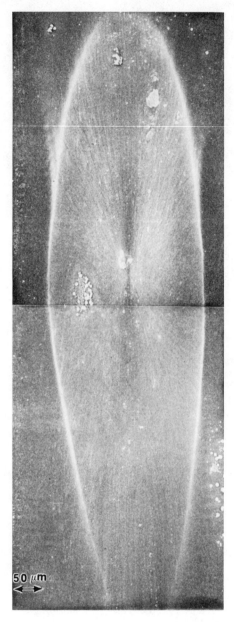

FIGURE 9. Isolated spherulite in an amorphous matrix crystallized at 160°C which has undergone uniaxial extension at room temperature. Extension rate was 100%/min across a sample gauge length of 1cm. Ultimate draw extension was approximately 500%. Magnification 200x.

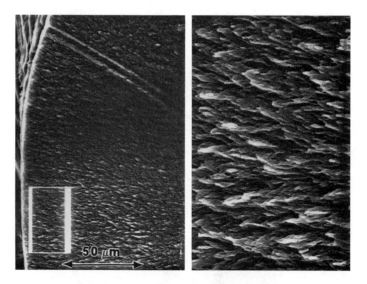

FIGURE 10a. Uniaxially drawn spherulite after standard etch
treatment. Conditions of draw are identical to those of Figure 9.
Split image highlights the splaying of crystalline fibrils in the
equatorial zone. Magnification 500x, insert magnification 2500x.

FIGURE 10b. Uniaxially drawn spherulite after standard etch
treatment. Conditions of draw are idential to those of Figure 9.
Split image highlights the polar regions of the spherulite where
plastic deformation of the crystalline fibrils may have resulted in
some melt recrystallization. Magnification 500x, insert
magnification 2500x.

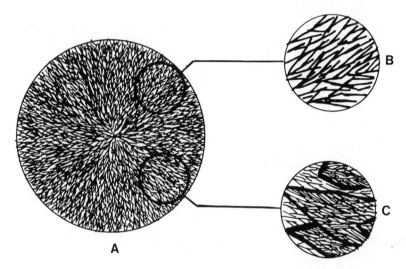

FIGURE 11. Schematic model of uniaxial spherulitic deformation.

A) General undeformed crystalline fibril structure remaining after course etching.

B) Finely branched structure observed in the enlarged region enhanced by fine etching but which is not apparent with courser etching conditions.

C) Typical spherulitic shape following uniaxial deformation – lines represent the general pattern followed by the courser crystalline fibril texture.

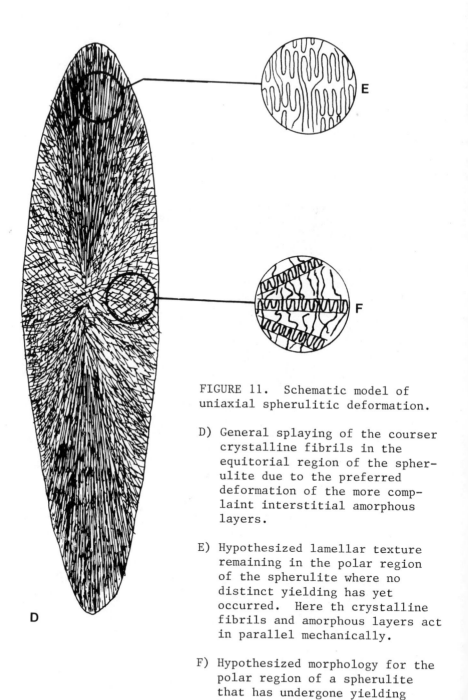

FIGURE 11. Schematic model of uniaxial spherulitic deformation.

D) General splaying of the courser crystalline fibrils in the equitorial region of the spherulite due to the preferred deformation of the more complaint interstitial amorphous layers.

E) Hypothesized lamellar texture remaining in the polar region of the spherulite where no distinct yielding has yet occurred. Here th crystalline fibrils and amorphous layers act in parallel mechanically.

F) Hypothesized morphology for the polar region of a spherulite that has undergone yielding based on a possible melting – recrystallization mechanism during yielding.

new and similar models have been suggested by others[11,12]. Specifically it is noted that when the crystalline fibrils and interstitial amorphous layers are lying perpendicular to the principal applied stress, these two elements are in fact basically acting in "series" from a mechanical point of view. Hence the lower modulus characteristics of the amorphous layers more easily undergo deformation while the crystalline fibrils that retain a higher modulus tend to resist deformation. Hence there is a splaying of the crystalline fibrils such that the amorphous material becomes more exposed and is more susceptable to etching behavior. Clearly this is noted from the micrographs where one observes that in the equatorial regions of the spherulite, enhancement of etching has occured. However, in the polar regions of the spherulite, etching does not necessarily provide good detail particularly directly along the polar areas of the spherulite. Here, of course, the crystalline fibrils and amorphous layers act more in a "parallel" arrangement with respect to deformation and hence the crystalline fibrils must begin to undergo plastic deformation if strain is to occur. In fact the deformation of the spherulite is not quite uniform if one compares the spherulite elongation above and below the spherulite center. We believe that the inhomogeniety of deformation is linked to the progression of the drawing neck through the spherulite. A systematic change in the nonuniformity is seen as a function of the draw condition. We expect to discuss this effect in full within a more detailed paper.

Biaxial deformation of spherulitic films was also investigated. Figure 12 shows an isolated spherulite which has undergone a 50% (approx.) biaxial strain at the conditions indicated within the figure caption. In this case the large spherulite is immersed within a matrix of much smaller spherulite induced through the second crystallization step described in the experimental section. Due to the fact that there is a much higher nucleation density generated by crystallizing from the glass, the matrix spherulites are small and difficult to observe at the magnification level of the micrograph shown. It was necessary to both provide a matrix of compatable mechanical characteristics and to raise the temperature well above Tg to effect the required biaxial spherulitic deformation. Apparently since all the stresses in biaxial deformation are borne by the spherulite along the radial directions in a circularly symetric manner, the deformation tends also to be circularly symetric. One observes concentric rings within the spherulite which represent alternately deformed and undeformed regions. This concentric ring necking phenomenon is more clearly observed in Figure 13 where the biaxially drawn spherulite has been chemically etched prior to SEM observation. The raised areas are undeformed regions (relatively) and show the clear radial

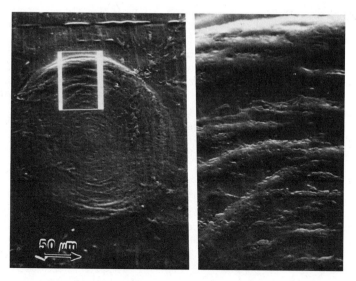

FIGURE 12. Isolated spherulite crystallized at 160°C in a matrix
of smaller volume filling spherulites subjected to a balanced
biaxial deformation at 140°C. Extension rate was 2000%/min across
5 cm square film sample. Final extension was approximately 50%.
Magnification 300x, insert magnification 1500x.

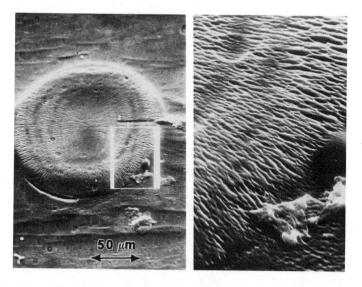

FIGURE 13. Biaxiially drawn spherulite after standard etch treatment. Conditions of draw were identical to those of Figure 12.
Split image highlights the concentration ring necking pattern within the spherulite. Magnification 300x, insert magnification 1500x.

structure of the spherulites well intact. The interstitial depressed areas are the drawn regions and show a much finer crystalline structure where it is assumed that the lamellar structure of the spherulite has been transformed into a much more oriented structure. This deformed structure is as yet uncharacterized. To the authors knowledge, this micrograph is one of the first to clearly relay detail of spherulite deformation under equal biaxial orientation.

With regard to biomaterial considerations we wish to point out that as a result of a deformation process, uniformity is not likely to occur within a spherulitic texture, at least at the low levels of strain reported here. This can often promote variations in orientation and residual stresses which will in turn lead to differences in transport characteristics. As evidenced by the dependence of hydrolysis rate upon morphology and deformation as described in this paper. One could also apply these same considerations to the transport of specific drugs through such a matrix.

SUMMARY

In conclusion, we believe that through the use of the hydrolysis method that we have applied to this specific copolyester the effect of some of the important crystallization parameters on morphological texture has been demonstrated. We also have tried to convey the message that morphology can be controlled or certainly varied by using specific thermal procedures but that this morphology can also be altered through deformation procedures which in turn will undoubtedly influence final properties.

REFERENCES

1. R. L. Kronenthal, in "Polymers in Medicine and Surgery," R. L. Kronenthal, Z. Oser, and E. Martin editors, Polymer Science and Technology, Vol. 8, Plenum Press, New York, 1979.

2. Bernard Wunderlich, "Macromolecular Physics, Vol. 1, Crystal Structure, Morphology, Defects". Academic Press, New York (1973).

3. S. Y. Hobbs, J. Macromol. Sci. Rev. Macromol. Chem., C19(2), 221, 265 (1980).

4. A. Keller and S. Sawada, Macromol. Chem., 74, 174 (1964).

5. O. A. Battista, Am. Sci., 53, 151 (1965).

6. O. A. Battista, M. M. Cruz, and C. F. Ferraro, in "Surface and Colloid Science," Vol. 3, (E. Matijeric, ed.), Wiley, New York (1971).

7. E. W. Fischer, H. J. Sterzel, and G. Wegner, Kolloid Z.Z. Polym., 251, 980 (1973).

8. S. W. Rowe and R. Tobazeon, J. Mat. Sci., 16, 2605 (1981).

9. H. D. Keith, F. J. Padden, Jr., and R. G. Vadimsky, J. Appl. Phys., 24(12), 4585 (1971).

10. H. D. Keith and F. J. Padden, Jr., J. Appl. Phys., 34, 2409 (1964).

11. H. D. Keith and F. J. Padden, Jr., J. Appl. Phys., 35(4), 1270 (1981).

12. Robert J. Samuels, "Structural Polymer Properties," Wiley, New York (1974).

13. Kaoru Shimamura, Syozo Murakami, Masaki Tsuji, and Ken-ichi Katayama, J. Soc. Rheol. (Japan), 7(1), 42 (1979).

STRUCTURAL IDENTIFICATION OF CIS-PLATINUM II POLYHYDRAZINES

Charles E. Carraher, Jr.,[a] Tushar A. Manek,[a] George
G. Hess,[a] David J. Giron,[b] Mary L. Trombley[a,b]
and Raymond J. Linville[a]

Departments of Chemistry[a] and Microbiology and Immunology[b]
Wright State University
Dayton, Ohio 45435

INTRODUCTION

Malignant neoplasms are the second leading cause of death
in the USA. Prior to the work of Rosenberg, only totally organic
compounds were investigated as possible antineoplastic agents.
Rosenberg (1) in 1964 found that bacteria failed to divide,
but continued to grow giving filamentous cells in the presence
of platinum electrodes. The cause of this inhibition to cell
division was eventually traced to the presence of a small crack
in the electrode which permitted minute amounts (about 10 ppm)
of cis-dichlorodiamineplatinum II, (cis-DDP), trans-dichloro-
diamineplatinum II (trans-DDP) and cis-tetrachlorodiamineplatinum
IV (cis-TCP), the activity of cis-DDP being the greatest. Experimentation with cancer cell lines, live animals, and human subjects
followed, focusing on cis-DDP. It was found that cis-DDP is
active against a wide range of tumors in man (2) and animals
(3) leading to the licensing of cis-DDP under the name of Platinol.

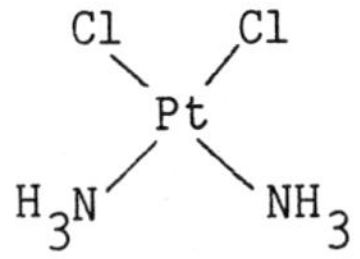

$$
\begin{array}{ccc}
Cl & & Cl \\
\diagdown & & \diagup \\
 & Pt & \\
\diagup & & \diagdown \\
H_3N & & NH_3
\end{array}
$$

1-cis-DDP

93

TOXICITY MINIMIZED

The positive attributes of platinum compounds are found
to be coupled with a number of negative side effects including
gastrointestinal, hematopoietic, immunosuppressive, auditory[3,4]
and renal dysfunction,[3,5,6] with the latter two being the most
serious.

A number of approaches have been employed to reduce the
toxicity of cis-DDP and its derivatives. The initial effort
was the hydration technique of Cvitokovic et al.[7] where the
patient's fluid uptake is greatly increased prior to the administration of the cis-DDP. The cis-DDP is then administered along
with mannitol, a diuretic. This technique allowed the administration of about ten times the normal dosage, while lowering damage
to the kidneys. It also signals a real problem related to retention of the drug. Typically over 80% of the water soluble cis-
DDP is flushed through the body and excreted within 16 hours
after administration and it is this high dosage of cis-DDP to
the body's circulatory system that is responsible for the majority
of negative biological effects.

Another early approach, which met with moderate success,
involved repeated application of the drug in an attempt to build
up the amount of platinum compound retained. It is believed
that small amounts of cis-DDP and its derivatives become associated with the body's protein and other cellular components where
the retained platinum has a body half-life in excess of a week.

Most of the more recent studies involve two approaches.
First, use of cis-DDP in conjunction with other drugs as adria-
mycin, vinblastine, bleomycin, actinomycin D and cyclophosphamide.
Second, the synthesis of new compounds exhibiting equal or enhanced
activity and lowered toxicity. This latter approach has been
taken by the research groups associated with Allcock[8-12] and
Carraher[13-21] with the inclusion of platinum diamines in polymers.

Most of the current efforts related to the synthesis of
platinum compounds for antitumor activity are aimed at new compounds exhibiting antitumor activity comparable to cis-DDP but
with decreased toxicity or greater therapeutic properties allowing
decreased dosage level. Such efforts have produced a wide variety
of compounds, a number of which show antitumor activity.[22] From
studying the structural features of such compounds several features
have emerged. Active compounds are a. neutral, b. contain
two inert and two labile ligands and c. must have the ligands
cis to each other.

Fortunately nature has provided a ready synthetic route
to the synthesis of both the cis of trans isomers at the exclusion

of the other isomer. The trans effect is in effect for many
Group VIII B square planer compounds, including Pt, Pd and Ni.
Thus reaction of tetrahaloplatinum II salts with nitrogen containing compounds gives exclusively the cis-isomeric product. The
trans-isomeric product (3) is obtained from reaction of the
tetraaminoplatinum II with halogens.

STRUCTURAL REQUIREMENTS

    Where both the cis and trans compounds were tested, the
cis isomers are found to be more reactive. The trans effect
also predicts that a chloride ligand is more readily replaced
(more labile) than the cis isomer. Thus trans-DDP hydrates
about four times faster than the cis isomer[23] and undergoes
ammination ten times faster.[24] This greater reactivity might
imply a lowered reaction specificity for the trans isomer.
Thus, even though the two isomers are of approximately equal
toxicity, their therapeutic levels differ vastly.[24-26] This
difference might be due to the reaction of the trans isomer
with various constituents of the body prior to reaching the
tumor site. Distribution and excretion studies showed cis-DDP
to be excreted much faster initially.[27] However, within five
days the levels were comparable at about 20% retention. Even
at this point the distribution of the two compounds differed
radically, platinum levels from trans-DDP remaining high in
plasma at all times, while levels from cis-DPP fall off markedly.[27,28] This might be explained by suggesting that trans-DDP
reacts with some constituents in the blood, remaining there
for some time, while cis-DDP reacts somewhat later, thus being
readily filtered from the blood by the kidneys.

The biological and chemical activity of the platinum drug is also dependent on the nature of the labile moieties. Thus, if the leaving groups are too reactive, the drug may chelate prior to reaching the tumor site. If the leaving groups are not labile enough, the drug will not chelate with the proper cellular material.

The nature of the amine ligand also affects the biological and chemical activity of the platinum drug.

First, aliphatic amines are found to offer better activity than aromatic amines.[26,27] Second, nonlabile amines are preferred. Third, solubility, hydrolysis rate and mobility are decreased as the organic portion of the amine is increased.[27,29]

A discussion of mechanistic pathways is reviewed in reference 30.

A point that is generally not appreciated regards the multiplicity of platinum-containing moieties once the cis-DDP, or other platinum-containing drug is administered. While it is known that the vast majority (>80%) of cis-DDP is readily flushed through the human body, the exact form of the stored, tumoral-active platinum-containing drug is not known. Thus, while there exists a great deal of "hand-waving" regarding the elements necessary to give a biologically acceptable platinum compound, the "established" structural requirements must be considered as first approximations and nonlimiting. For instance, the idea that the platinum drug should be neutral is counter to the experience of researchers dealing with charged assemblages of platinum-containing compounds called the "platinum blues". It is true that the platinum blues appear to inhibit tumoral growth through other routes than does cis-DDP but even this point is not sure.

RATIONALE

The use of polymers containing platinum compared with platinum delivery through use of smaller molecules is clearly debatable, but there is sufficient evidence to justify preliminary studies of such materials. The use of biological and synthetic polymers as tumor-suppressant is well known and has in some cases proven to be advantageous (for instance 31-34).

Specifically advantages for synthesizing polymeric derivatives of cis-DDP includes a. restricted biological movement, b. controlled release, c. increased probability of critical attachment and d. delivery of increased amounts of drug.

Heavy metal toxicity-related to the presence of large quantities of cis-DDP derivatives in the circulatory system (such
as renal failure), is well established. Chain lengths of about
100 and greater are typically prevented from easy movement through
biological membranes. Thus the location of the platinum drug
can be somewhat restricted, for instance from the kidney, decreasing damage to the kidney-associated organs.

Studies by one group have established that most metal-contain-
ing polymers undergo hydrolysis when wetted. DMSO solution
Polymeric derivatives of cis-DDP in DMSO solution also undergo
hydrolysis when added to water. Thus, polymeric derivatives
of cis-DDP can act as controlled release agents, releasing therapeutic quantities of the active drug.

If attachment - interstrand, intrastrand or otherwise -
to DNA is essential for antitumor activity and multiple attachments
are required or advantageous (i.e. more than one attached platinum
compound per DNA) then the fact that the platinum is itself
an integral portion of a polymer is advantageous since the probability that successive attachments will be made on the same
strand and adjacent strands is high after the first attachment.

An often employed upper dosage of cis-DDP (single dose)
is about 4×10^{-4} g/kg for humans which is increased to $4 \times
10^{-3}$ g/kg where flushing is employed.[7] Mice have been injected
(note that the cis-DDP is administered orally) on an alternate
day schedule for one month with DMSO-H_2O (10-90% by volume)
solutions containing about 200 times this amount (2×10^{-2} g/kg)
without apparent harm with various platinum II polyamines.[26]

We recently reported the synthesis of a number of polymeric
platinum II derivatives of cis-DDP.[13-21] A number of synthetic
routes were developed permitting the synthesis of the chloro,
bromo and iodio derivatives as described below.

where X = Cl,Br,I.

4

where Y = O,S.

5

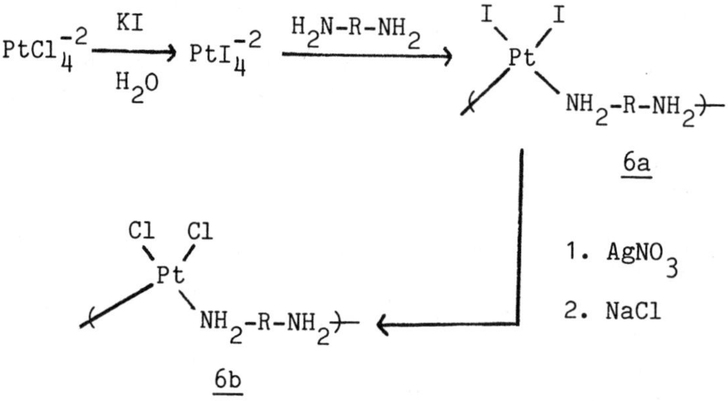

$$PtCl_4^{-2} \xrightarrow[H_2O]{KI} PtI_4^{-2} \xrightarrow{H_2N-R-NH_2}$$

6a

6b

1. AgNO$_3$

2. NaCl

The latter scheme deserves special mention. The reaction rate is of the order I > Br > Cl where good yields are obtained with PtI_4^{-2} after several minutes whereas reactions employing $PtCl_4^-$

require a day for decent (>50%) yields to occur. The above reaction sequence allows the synthesis of good yields of the chloro polymer within several hours.

The platinum II polyamines (where R ≠ aromatic) typically exhibit good inhibition towards a wide range of tumor cells including WISH, HeLa and L929 at concentrations within the range of 10 to 40 ug/ml. Some extend the life of (induced) terminally ill (cancerous) mice about 50%. Many exhibit good cell differentiation favoring inhibition of transform cells over analogous healthy cells. Recent biological studies are aimed at evaluation of select platinum II polyamines against specific viruses and tumors.

Recent chemical efforts are aimed at the generation of kinetic data (synthesis and biological degradation), inclusion of nitrogen-containing reactants which themselves exhibit antitumoral and antiviral activities and the extention of the reaction to include other nitrogen-containing reactants.

Here we report the initial synthesis and a brief chemical and biological characterization of platinum II polyhydrazines of the following form.

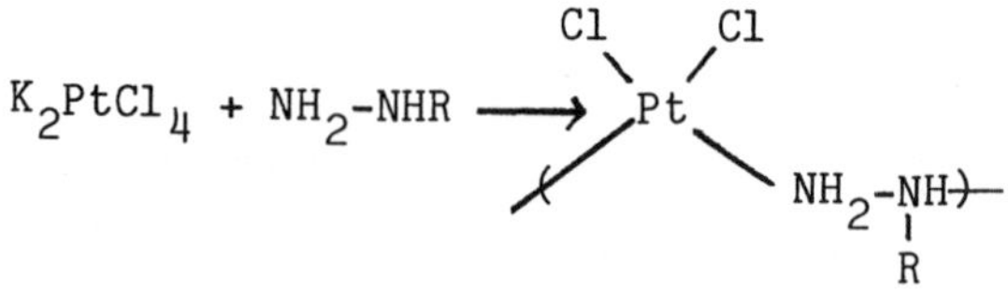

$$K_2PtCl_4 + NH_2-NHR \longrightarrow$$

7

The importance of such a study is illustrated by considering that a number of drugs contain hydrazine moieties with the synthesis of platinum II polyhydrazines derived from relatively inexpensive hydrazines assisting in determining possible reaction dependencies of the hydrazine moiety.

EXPERIMENTAL

Syntheses were accomplished employing aqueous solutions containing the platinum compound and the hydrazine in an open Kimex glass beaker employing constant stirring furnished by a magnetic stirrer. The platinum II polyhydrazine precipitates from the reaction mixture and is removed by suction filtration and washed repeatedly with ionized, doubly distilled water (employed throughout). The polyhydrazine is washed with water into a preweighed glass petri dish and permitted to dry in air. The product is typically a light to dark brown powder.

Infrared spectra were obtained using KBr pellets employing Perkin-Elmer 457 and 1330 Infrared Spectrophotometers and a Digilab FTS-20 C/D FT-IR. The latter has a resolution of 0.1 cm^{-1} and is equipped with a digitally controlled plotter, a Spectramate graphic and alphanumeric operator terminal and related hardware and software.

Weight-average molecular weights were obtained using a Brice-Phoenix 3000 Ligh-Scattering Photometer employing a Bausch and Lomb Refractometer (Model Abbe-31).

Solubilities were attempted in a wide variety of solvents including the dipolar aprotic solvents DMSO, DMF, HMPA, acetone and TEP by placing about 1 mg of product in with 3 ml of liquid and observation for about a week.

The following reagents were employed as received: potassium tetrachloroplatinum (II) (J and J Materials), phenylhydrazine (Baker Chemical Co.,), p-tolylhydrazine (Aldrich), 4-bromophenylhydrazine (Eastman), 4-fluorophenylhydrazine (Aldrich), 4-chlorophenylhydrazine (Aldrich), p-thiocyanophenylhydrazine (Alfred Baber-Aldrich), 4-nitrophenylhydrazine (Aldrich), and 4-methylsulfonylphenyl hydrazine (Aldrich).

Mass spectral analyses were performed using a direct insertion probe in a Kratos MS-50 mass spectrometer, operating in the EI mode, 8 KV acceleration and 10 sec/decade scan rate. UV-VIS-NIR spectra were obtained in DMSO employing a Cary 14 spectrophotometer. NMR spectra were obtained in d$_6$-DMSO using a Varian EM360A spectrometer. Elemental analyses were accomplished employing a Perkin-Elmer 240 Elemental Analyzer for C, H, N and through

Bacterial tests were conducted in the usual fashion. Tryptic soy agar plates were seeded with suspensions of the test organism to produce an acceptable lawn of test organism after 24 hours of incubation. Shortly after the plates were seeded, the tested compounds (0.1 mg) were deposited as solids on the plates as small round spots. The plates were incubated and inhibition noted.

Two cell lines were tested, L929, a mouse connective tissue tumor cell line and HeLa, a human cervical cancer cell line. The cell lines were trypsinized, cells suspended and counted, then diluted to 10^6 and plated onto Corning 1 ml well plates. The plates were incubated overnight in a 100% humidity carbon dioxide incubator and the next day the DMEM sucked off and substituted with DMEM containing various microgram quantities of the platinum-containing polymer. Inhibition was tested for by both visual observation and employing trypan blue (an exclusion stain).

RESULTS AND DISCUSSION

The physical characterizations of the products are consistent with a repeat unit as depicted by form 7. Selected results follow.

Structural evidence is derived from control reactions, historical arguments, elemental analyses, spectral data (mainly UV-VIS, MS and IR) and molecular weight determinations. Historically, as noted in the introductory section, reaction with nitrogen-containing compounds results in the formation (chelation) of Pt-N compounds, through displacement of two chloride atoms, with the ligands cis to one another. Control reactions, where one of the reactants is omitted, were run and no precipitate formed consistent with the product containing moieties derived from both reactants.

The ultraviolet spectra of the platinum II polyhydrazines shows three strong peaks roughly corresponding to spectra of other cis-platinum II compounds and less similar to spectra of the trans-DDP consistent with a product of form 7 (Table 1).

Infrared spectra are similar to the spectra of the nitrogen-containing reactant. In fact, it is known that the platinum has little effect on the bond strengths of amines, other than to weaken the N-H stretching vibrations.[35-37] Differences are noted in the 420 to 550 cm^{-1} region where two bands attributed to the Pt-N stretching mode are found and the presence of a single new band in the 300 to 350 cm^{-1} region attributed to the Pt-Cl stretching mode.

Table 1. UV bands of various platinum compounds

	Wavelength[a]		
C-DDP	240	273	331
t-DDP	-	268	317
C-HMDA-PtCl$_4$[b]	250	310	340
p-Bromophenylhydrazine, PtCl$_4{}^{-2}$	275	290	340
Phenylhydrazine, PtCl$_4{}^{-2}$	265	289	339

[a]All wavelengths in nm's.

[b]Product from K$_2$PtCl$_4$ and hexamethylenediamine.

Infrared spectra contain bands characteristic of the presence of both reactants and the presence of new bands characteristic of the formation of the Pt-N moiety. For the product derived from 4-nitrophenylhydrazine the following (selected) bands are present: (all bands given in cm^{-1}) broad band between 3600-3050 characteristic of N-H stretching; 1570 N-H deformation, 1500 C-NO$_2$ assymetric stretching; 1340 and 1290 C-NO$_2$ symmetric and C-N (aromatic) stretching, O-N=O stretching at 755; 490 and 470 (cis) Pt-N stretching and 323 Pt-Cl stretching (Figure 1).

Table 2 contains a summation of bands associated with the Pt-N and Pt-Cl stretching moieties. The number of Pt-N stretching vibration bands is often taken as evidence of the geometry being cis or trans with trans platinum amines showing one Pt-N band while the presence of two bands indicates a cis geometry.[35,37]

Mass spectral results for two compounds are given in Tables 3 and 4. All compounds scanned were above m/e 50.

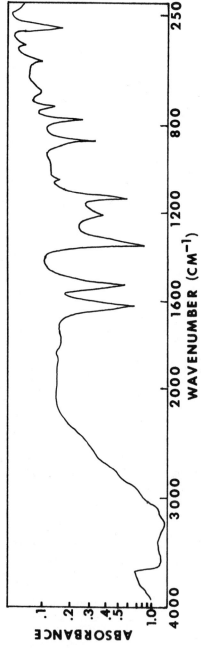

Figure 1. Infrared spectrum of the product derived from K_2PtCl_4 and p-nitrophenylhydrazine.

Table 2. Location of Specific IR Bands for Products of $PtCl_4^{-2}$ and Derivatives of p-Phenylene Hydrazine

Para-Substituent	Pt-N	Pt-Cl
NO_2	490, 470	323
H	495, 470	340
Br	480, 470	310
CH_3	470, 455	310
H_3CSO_2	480, 455	310

In Table 3 are listed the major ions in the mass spectrum of the product derived from K_2PtCl_4 and p-nitrophenylhydrazine. The formulas of the ions listed in column 3 are supported by the high resolution mass measurement to within the deviation shown in column 4. The structural formulas in column 5 are suggestions based on the structure of the initial molecule. The mass spectrum of the polymer differs from that of pure p-nitrophenylhydrazine. For the latter, the molecular ion (m/e 153) is the base peak. All of its fragment masses are present in the polymer spectrum, although the coresponding masses have greater intensity relative to m/e 153, in the spectrum of the polymer than in the pure monomer.

In Table 4 are listed all the ions of the p-bromophenyl-hydrazine polymer above m/e 260, and the major ions below that value. Identifications in this system are based on the bromine isotope ratios rather than the high resolution mass measurements. It is apparent that during the thermal decomposition, reactions of two or more p-bromophenylhydrazine groups with each other, accompanied by rearrangements and/or partial, decomposition have occurred. No peaks corresponding to the molecular ion, $C_6H_7BrN_2$, m/e 186 and 188, were observed, although weak 183 and 185 were present.

Thus the mass spectra are consistent with a product containing the hydrazine moiety.

Table 3. Mass Spectrum of Dichloroplatinum(II) Poly-(p-nitro-phenylhydrazineplatinum(II) dichloride.

Nominal m/e	Rel. Int. (%)	Formula	Mass Dev. (ppm)	Possible Structure
153	11.29	$C_6H_7N_3O_2$	2.7	$NH_2NH-C_6H_4-NO_2^+$
138	12.24	$C_6H_6N_2O_2$	-0.3	$NH_2-C_6H_4-NO_2^+$
123	32.70	$C_6H_5NO_2$	-3.5	$C_6H_5-NO_2^+$
123	7.33	$C_6H_7N_2O$	0.4	$ON-C_6H_4-NH_3^+$
108	68.77	C_6H_6NO	2.5	$C_6H_5-NOH^+$
92	11.72	C_6H_6N	-0.5	$C_6H_5-NH^+$
77	36.14	C_6H_5	15.3	$C_6H_5^+$
67	12.99	$C_5H_7^+$	-3.4	
65	100.00	$C_5H_5^+$	-16.7	
64	27.59	$C_5H_4^+$	-6.4	

The polymeric nature of the products is shown by the molecular weight determinations employing light scattering photometry (Table 5). This does not eliminate the formation of six-membered rings as depicted below since similar products are water soluble and our product is isolated as a precipitant from aqueous solution. Thus the longer chains may be formed directly or through such cyclic intermediates. This is currently under study.

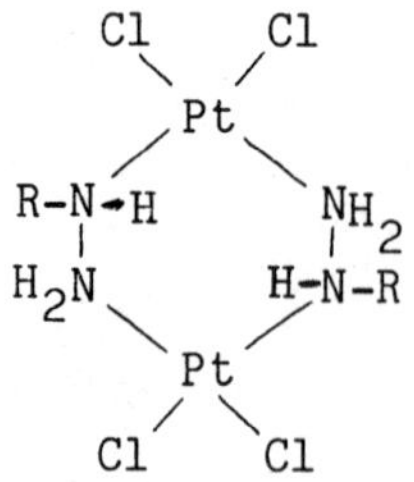

Table 4. Mass Spectrum of Dichloroplatinum(II) Poly-(p-bromophenylhydrazineplatinum(II) dichloride.

m/e	Rel. Int. (%)	Possible Structure
342	2.26	$Br-C_6H_4-N_2-C_6H_4-Br^+$
340	4.25	
338	2.43	
330	1.39	
329	1.40	
328	2.87	
327	2.86	
326	1.37	
325	1.12	
314	5.14	$Br-C_6H_4-C_6H_4-Br^+$
312	10.52	
310	5.81	
238	4.62	$Br-C_6H_4-Br^+$
236	10.28	
234	5.33	
185	3.29	$Br-C_6H_4-N_2^+$
183	3.80	
158	18.05	$C_6H_5Br^+$
156	17.38	
155	10.63	
83	10.28	HBr^+
82	100.00	Br^+
81	52.90	HBr^+
80	93.96	
78	6.56	
77	9.70	
73	9.59	
69	19.32	
68	5.50	
67	13.05	
57	13.05	
55	26.47	

Table 5. Results as a Function of Hydrazine.

Hydrazine	Yield (%)	M_w (10^{-5})	$\overline{DP}$
Phenylhydrazine	14		
4-Fluorophenylhydrazine	23		
4-Chlorophenylhydrazine	28	2.3	605
4-Bromophenylhydrazine	39		
p-Thiocyanophenylhydrazine	61		
4-Nitrophenylhydrazine	64[a]		
4-Methylphenylhydrazine		12.7	3,000
4-Methylsulfonylphenylhydrazine	37	1.6	420

Reaction conditions: K_2PtCl_4 (2.00 mmoles) in 5 ml water added to stirred solutions containing the hydrazine (2.00 mmole) in 20 ml for 24 hours stirring time.

[a] for 3.5 hours.

The reaction is general with decent yields obtainable after several hours to a day. Reactions employing thiourea and thioamide reactive groups require several days for decent yields to occur while the analogous urea and amide compounds require a week or longer before decent yields are obtained.[17] Thus the general reactivity decreases as the electron density on the nitrogen decreases consistent with what would be expected for simple ligand exchange.

Preliminary biological testing was accomplished. The products (as solids) were tested against a number of divergent bacteria (Escherichia coli, Alcaligenes faecalis, Staphylococcus epidermis, Staphylococcus aureus, Enterobacter aerogenes, Neisseria mucosa, Klebsiella pneumoniae, Psuedomonas aeroginosa, Actinobacter calcoaceticus and Branhamella catarohilis) and were largely noninhibitory with the methylsulfonylphenylhydrazine inhibiting only N. mucosa, P. aeroginosa and B. catarohilis.

Two tumoral cell lines were tested to determine the inception of inhibition. For many platinum II polyamines inhibition begins within the range of 6 to 20 ug/ml with complete inhibition typically occurring from 20 to 50 ug/ml. Inhibition for the platinum II polyamines is within this range (Table 6). Thus preliminary results indicate that further testing is merited.

Table 6. Inhibition of Platinum II Hydrazines on Tumor Cells.

Phenyl Hydrazine	Concentration (ug/ml)						
	12 L929	10	8	6	5	4	3
4-Methylsulfonyl				1	0	0	0
4-Nitro	4	2	2	2	2	1	1
4-Bromo				0	0	0	0
4-Chloro				0	0	0	0
4-Fluoro				0	0	0	0
Phenolhydrazine	4	2	1	0		0	
	HeLa						
4-Methylsulfonyl							
4-Nitro	4	2	1	1	0	0	
4-Bromo				0	0	0	
4-Chloro	3	2	1	0	0	0	
4-Fluoro				1	0	0	
Phenylhydrazine	4	2	1	0			0

where 0 = no inhibition,...,4 = 100 inhibition

REFERENCES

1. B. Rosenberg, L. Van Camp and T. Krigas, Nature (London), 205, 698 (1965).
2. J. Gottlieb and B. Drewinko, Cancer Chemother. Rep., Pt.1, 59, 621 (1975).
3. B. Rosenberg, Cancer Chemother. Rep., Pt.1, 59, 589 (1975).
4. S. Stadnicki, R. Fleischman, U. Schaeppi and P. Merriman, Cancer Chemother. Rep., Pt.1, 59, 467 (1975).
5. J. Ward, D. Young, K. Fauvie, M. Wolpert, R. Davis and A. Guarino, Cancer Treatment. Rep., 60, 1675 (1976).
6. J. Ward and K. Fauvie, Tox. and Appl. Pharm., 38, 535 (1976).
7. D. Hayes, E. Cvitokovic, R. Golby, E. Scheiner and I. Krakoff, Proceedings of American Assoc. Cancer Res., 17, 169 (1976).
8. H. Allcock, R. Allen and J. O'Brien, Chem. Comm., 717 (1976).
9. H. Allcock, Science, 193, 1214 (1976).

10. H. Allcock, Polymer Preprints, 18, 857 (1977).
11. H. Allcock, Organometallic Polymers (C. Carraher, J. Sheats and C. Pittman, Editors), Academic Press, N.Y. (1978), pgs. 283-288.
12. H. Allcock, R. Allen and J. O'Brien, J. Amer. Chem. Soc., 99, 3984 (1977).
13. C. Carraher, C. Admu-John and J. Fortman, unpublished results.
14. C. Carraher, D.J. Giron, I. Lopez, D.R. Cerutis and W.J. Scott, Organic Coatings and Plastics Chemistry, 44, 120 (1981).
15. C. Carraher, W.J. Scott, J.A. Schroeder and D.J. Giron, J. Macromol. Sci.-Chem.A15(4), 625 (1981).
16. C. Carraher, Organic Coatings and Plastics Chemistry, 42, 428 (1980).
17. C. Carraher, T. Manek, D. Giron, D.R. Cerutis and M. Trombley, Polymer Preprints, 23(2), 77 (1982).
18. C. Carraher and A. Gasper, Polymer Preprints, 23(2), 75 (1982).
19. C. Carraher, Biomedical and Dental Applications of Polymers (G. Gebelein and F. Koblitz Editors), Plenum Press, N.Y., 1981, Chpt. 16.
20. C. Carraher, M. Trombley, I. Lopez, D.J. Giron, T. Manek and D. Blair, unpublished results.
21. C. Carraher, H.M. Molloy, M.L. Taylor, T.O. Tiernan and W.J. Scott, unpublished results.
22. R. Speer, H. Ridgway, L. Hall, D. Stewart, K. Howe, D. Lieberman, D.A. Newman and J. Hill, Cancer Chemother. Rep., Pt. 1, 59 629 (1975).
23. M. Tucker, C. Colvin and D. Martin, Inorganic Chem., 3, 1373 (1964).
24. C. Colvin, R. Gunther, L. Hunter, J. McLean, M. Tucker and D. Martin, Inorganic Chimica Acta, 3, 487 (1968).
25. T. Conners, M. Jones, W. Ross, P. Braddock, A. Khokharard, M. Tobe, Chemico-Biological Interactions, 5, 415 (1972).
26. M. Cleare and J. Hoeschele, Platinum Metals Rev., 17, 2 (1973).
27. M. Cleare and J. Hoeschele, Bioinorg. Chem. 2, 187 (1973).
28. C. Litterset, T. Gram, R. Dedrick, A. Leroy and A. Guarino, Cancer Res., 36, 2340 (1966).
29. M. Tobe and A. Khokhar, J. Clinical Hematology and Oncology, 7, 114 (1977).
30. C. Carraher, W.J. Scott and D.J. Giron, "Bioactive Polymers" (Edited by C. Geberlein and C. Carraher), Plenum, 1983.
31. C. Carraher and C. Gebelein, Eds., "Bioactive Polymers," Plenum, 1983.
32. G. Rowland, G. O'Neil and D. Davis, Nature, 255, 487 (1975).
33. H. Ringsdorf, Middland Macromolecules Meeting (Edited by H. Elias), Marcel Dekker, N.Y., 1978.
34. H.J. Ryser, Nature, 215, 934 (1967).

35. A.A. Grinberg, M. Serator and M.I. Gel'fman, Russian J. Inorg. Chem., <u>13</u>, 1695 (1968).
36. R. Silverstein, G. Bassler and T. Morrill, "Spectronic Identification of Organic Compounds," Third Edition, John Wiley, N.Y., 1974, pages 107-109.
37. G. Barrow, R. Krueger and F. Basole, J. Inorganic and Nuclear Chemistry, <u>2</u>, 340 (1956).

Acknowledgement: The mass spectra were obtained at the Midwest Center for Mass Spectrometry, supported by NSF Regional Instrumentation Facility Grant #CHE 78-18572.

FLUORESCENCE IN POLYMERS: 2-DIPHENYLACETYL-1,3-INDANEDIONE-1-IMINE

DERIVATIVES IN POLYMER MATRICES

Catherine A. Byrne
Polymer Research Division, Organic Materials Laboratory
Army Materials and Mechanics Research Center
Watertown, Massachusetts 02172

Edward J. Poziomek
Research Division, Chemical Research and Development
Center, Aberdeen Proving Ground, Maryland 21010

Orna I. Kutai, Steven L. Suib and Samuel J. Huang
Department of Chemistry and Institute of Materials
Science, University of Connecticut, Storrs, Connecticut
06268

INTRODUCTION

Applications of fluorescence effects in polymers, both in solution or in solid polymer matrices are found in the literature, many related to chemical or physical changes in the polymers[1]. The purpose of this investigation is to prepare a fluorescent polymer system in which fluorescence enhancement can be used as a probe of chemical environment. Two fluorescent dyes, 2-diphenyl-acetyl-1,3-indanedione-1-imine derivatives, have been chosen for the study. These dyes have been shown to complex with many chemical compounds on solid supports such as glass fiber paper and exhibit a strong fluorescence enhancement[2,3]. Enhancement is observed with some insecticides, herbicides, α-hydroxy acid esters and α-amino acids. Of the fluorescent compounds tested, 3 in equation (1) exhibited the strongest effect. This is a report of the preparation of polyurethane elastomers containing 3 and the hydroxyethyl analog of 3,6 in Scheme 1, and the fluorescent properties of these polymers.

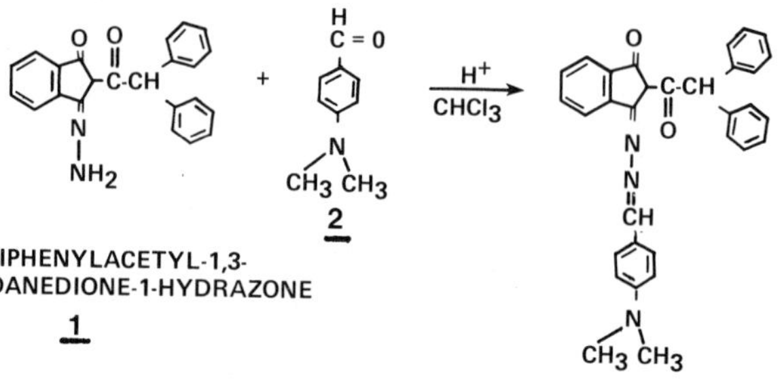

$$\text{(1)}$$

2-DIPHENYLACETYL-1,3-
INDANEDIONE-1-HYDRAZONE

1

2

2-DIPHENYLACETYL-1,3-INDANEDIONE-1-
(P-DIMETHYLAMINOBENZALDAZINE)

3

EXPERIMENTAL

Synthesis of Fluorescent Dyes

Red-orange azine 3 in equation (1) was prepared from hydrazone 1 (Aldrich) and p-dimethylaminobenzaldehyde according to
Braun and Mosher[4] using a few drops of concentrated HCl as
catalyst.

The ρ-bis(2-hydroxyethyl)aminobenzaldehyde 4 in Scheme 1 was
prepared by the method of Sen & Price[5]. It was obtained as an oil
which was estimated to be 50 percent pure by high performance
liquid chromatography (HPLC). Several attempts to purify the
material by preparative LC were unsuccessful, although the effort
was not exhaustive.

The azine 6 was prepared from the aldehyde 5 and hydrazone 1.
In 50 ml ethanol were mixed 5 g 1 (0.014 mole), 5.3 g 5 (0.013
mole assuming that the oil is 50 percent 5) and 5 drops piperidine. The mixture was refluxed for 20 minutes and filtered hot.
The brownish red solid product in the filter paper weighed 6.12 g
(80.2%). The low solubility of the product in recrystallizing
solvents necessitated the use of column chromatography to purify
6. A slurry of 2 g reaction product and silica gel in tetrahydrofuran was dried and placed atop a silica gel column. Elution with
1-5% ethanol in methylene chloride yielded 0.4 g 6 (20%), mp
262-265°C (DSC).

Anal. Calcd. for: $C_{34}H_{31}N_3O_4$: C, 74.86; H, 5.69; N, 7.71.
Found: C, 74.63; H, 5.66; N, 7.63.

An earlier reported procedure[6,7] using HCl as catalyst resulted in a similar compound with a slightly lower melting point,
a virtually identical infrared spectrum, a carbon-13 NMR which

112

SCHEME 1

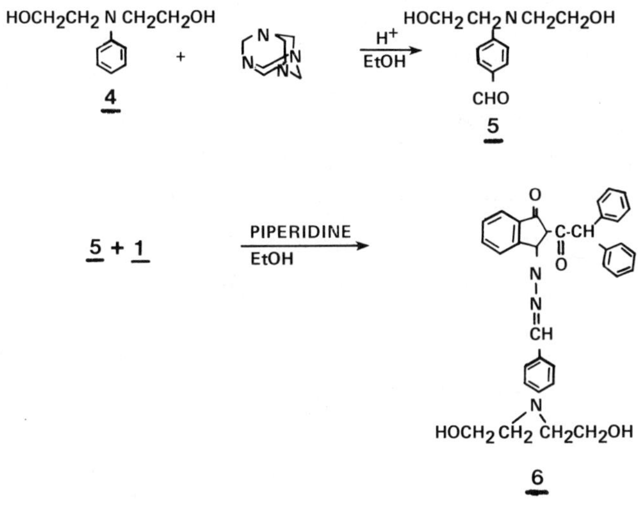

would be expected for 6, but an analysis which was low in carbon.
The difference between the two compounds has not been clarified.
All of the work described here involves the use of the compound
prepared with piperidine.

It should be noted that 3 and 6 have the potential for
several different types of resonance structures involving enol
formation[4]. Since the dyes are added to polyurethane reaction
mixtures, there is the possibility that the enol has reacted with
an isocyanate. A small quantity of 3 and an excess of p-chloro-
phenyl isocyanate were refluxed in dioxane for one hour. The
mixture was poured into methanol and analyzed by HPLC. It was
estimated that the enol had reacted with the isocyanate to the
extent of five to ten percent under these relatively harsh
conditions. This does not affect the results described semi-
quantitatively later in this paper.

Polyurethane Synthesis

The polyurethane elastomers were prepared using the prepoly-
mer technique shown in Scheme 2 and four weight-volume percent
dibutyl tin dilaurate (T-12, M&T Chemical) in methyl ethyl ketone
as catalyst. The synthesis has been described in detail[8]. All
the samples used in this work are composed of 29.2 weight percent
hard segment (hydrogenated methylene dianiline diisocyanate plus
1,4-butanediol). The isocyanate to hydroxy group ratio (NCO/OH)
is equal to 1.05. The fluorescent dyes were added in tetrahydrofuran solution at the chain extension step, with 1,4-butanediol.

113

SCHEME 2

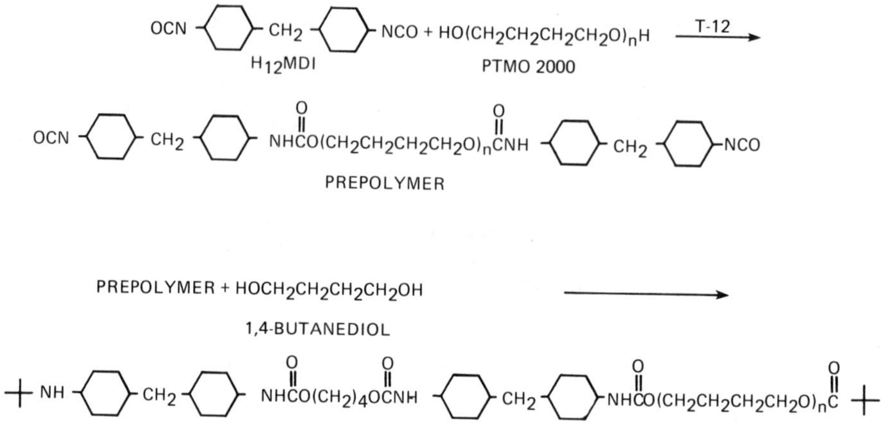

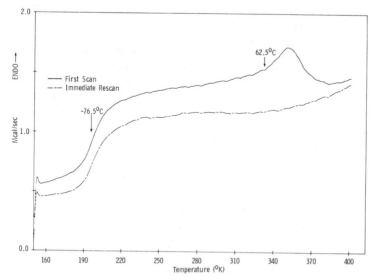

The polymers were prepared as 50 mil sheets in polypropylene molds or were cured in polystyrene or glass cuvettes for the spectroscopy experiments.

A DSC thermogram for one of the polyurethane samples is shown in Figure 1. The thermal properties of the polymers containing small amounts of the dyes are undistinguishable from those of the plain polyurethanes by DSC. The samples are phase segregated, having soft segment glass transition temperatures of about −77°C

Figure 1. DSC scan of a polyurethane sheet containing 3.67×10^{-6} moles $\underline{3}$ [RN(CN$_3$)$_2$] per 100 g polymer. Sample weight 14.29 mg. Scan rate 20°C/min.

and endothermic peaks with onsets of about 63°C and maxima of 74°C. This endotherm has been attributed to a partial ordering of the hard segments[9]. The presence of three geometric isomers of methylene bis(4-cyclohexylisocyanate)(Desmodur W, Mobay) in the hard segment inhibits crystallization and the polyurethane is transparent[10].

Fluorescence Spectroscopy

The fluorescence spectra were obtained using either a Spex Fluorolog or a Perkin-Elmer LS-5 spectrophotometer. Further details of the spectroscopy will be given in the next section.

RESULTS AND DISCUSSION

The ultraviolet-visible spectrum of $\underline{3}$ (10^{-4}M) in chloroform

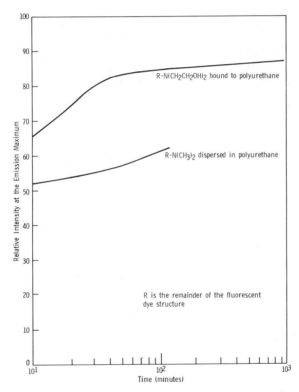

Figure 2. Fluorescence intensity as a function of cure time at 60°C. concentration of $\underline{3}$ [RN(CH$_3$)$_2$] or of $\underline{6}$ [RN(CH$_2$CH$_2$OH)$_2$] is 3.67×10^{-6} moles/100 g polymer. The excitation wavelength was 476 nm. The emission maximum occurs at about 565 nm for the polymer with $\underline{3}$ and at 555 nm for the polymer with $\underline{6}$. The Spex Fluorolog was used in the right angle mode.

exhibits two peaks, one at 340 nm and a maximum at 471 nm. The fluorescence spectrum exhibits a single peak with a maximum at 556 nm, which shifts slightly with a change in the concentration in solution. The fluorescence spectra of the dyes in the polyurethanes will be discussed below.

Progress of Polymerization

It has previously been demonstrated that the change in the fluorescence intensity of an added dye can be used to follow the course of polymerization[11,12]. In Figure 2 are compared the increase in fluorescence with polyurethane polymerization of 3 blended in the polymer matrix and 6 which is capable of covalent bonding in the polymer main chain. The reaction took place in the beam of the instrument at 60°C, a temperature which was chosen because the polystyrene cuvettes used could not withstand a higher temperature. In later work, glass cuvettes were adopted.

After three hours, the fluorescence for the sample containing 3 had increased 20% and for that containing 6 had increased about 30%. The maxima in the emission spectra exhibit hypsochromic shifts of about 10 nm during polymerization. Note that the fluorescence intensities of the two samples at the beginning of the

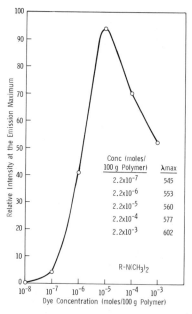

Conc (moles/ 100 g Polymer)	λmax
2.2×10^{-7}	545
2.2×10^{-6}	553
2.2×10^{-5}	560
2.2×10^{-4}	577
2.2×10^{-3}	602

R-N(CH₃)₂

Figure 3. Fluorescence as a function of concentration of 3 [RN(CH$_3$)$_2$] in the polyurethane. The Perkin-Elmer LS-5 was used in the front face mode. The polyurethane sheets were placed behind a quartz window. The excitation wavelength was 476 nm.

reaction cannot be compared directly because the maxima in the
excitation spectra differ by a few nanometers, but in this case
the same excitation wavelength was used for both. The shapes of
the curves in Figure 2 are noticeably different. The fluorescence
increases more rapidly in the upper curve, where the dye 6 is
incorporated into the polymer backbone in the hard segment, but
increases only gradually in the lower curve for the polymer where
the dye is only mixed in. Nonradiative fluorescence decay processes are expected to be inhibited when the dye becomes part of a
large polymer chain. At three hours reaction time, about 70% of
the isocyanate has reacted, determined by observation of the 2250
cm^{-1} infrared absorption.

Dye Concentration

In order to determine an appropriate concentration of the
fluorescent dye to use in the experiments, polymers were prepared
containing different concentrations of 3. The fluorescence intensities of the polymers are compared in Figure 3. The intensity
reaches a maximum at 10^{-5} moles/100 g polymer and then decreases
with increasing concentration of dye. This decrease may be due to
concentration quenching or to formation of aggregates of the dye
molecules which are less fluorescent. Concentrations in the 10^{-6}
moles/100 g ($\approx 10^{-5}$ M) region were convenient for this work because
the fluorescence was sufficiently intense and the likelihood of
complicating effects due to a too high concentration was reduced.
Note also the large shift in the emission maximum with an increase
in concentration which might be an indication of dye aggregation
in the polymers. The maxima in the excitation spectra exhibit
only a small shift of two to three in nanometers in the concentration range studied, which is within the instrumental error limit.

Fluorescent spectra for the polymers also exhibit a small
peak with a maximum slightly below 400 nm when excited at 343 nm
which decreases in size with increasing concentration of the dye.
This is due in part to the Raman scattering band of the polymer,
but also probably includes emission due to the dibutyl tin dilaurate catalyst and perhaps some trace fluorescent impurities present
in the starting materials for the polyurethane.

Effect of Temperature

The temperature at which the fluorescence of a sample is
measured generally has an effect on the fluorescence intensity.
The fluorescence intensity decreases uniformly with an increase in
temperature, but the behavior changes when a polymer sample passes
through a thermal transition. Loutfy[11] has observed a linear
decrease in fluorescence of an added dye with a temperature increase in poly(methyl methacrylate) until the glass transition
temperature is reached, above which the behavior is also linear,

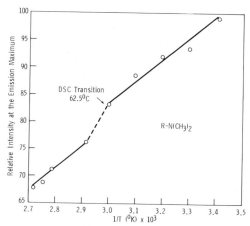

Figure 4. Fluorescence emission as a function of temperature for a sample polymerized in a glass cuvette containing 2.2×10^{-6} moles 3/100 g. The LS-5 equipped with a thermostated cell holder was used in the right angle mode, with excitation at 476 nm.

but the line exhibits a decrease in the absolute value of the slope. Isotactic poly(methyl methacrylate) exhibits an endotherm in the DSC at about 43°C which is not a glass transition, but rather is due to a change in polymer conformation, beginning with the motion of the methoxycarbonyl side groups[13]. The fluorescence behavior of an added dye in the latter polymer involves a linear decrease followed by a sharp drop in fluorescence at about 43°C, followed by linear behavior with the same slope as the initial slope. Similar behavior is observed in Figure 4 for the polyurethane. A sharp drop occurs at about 60°C, compared to the onset of the transition in the DSC of 62.5°C (Figure 1). This is interesting because the DSC transition occurs in a region where one might expect to find a T_g for the polyurethane. This fluorescence behavior corroborates DSC evidence (Figure 1) that since the transition does not reappear on rescan, it cannot be a glass transition.

Enhancement with Methyl Benzilate

A comparison of polyurethanes containing only 3 or 6 with those also containing methyl benzilate 7 is shown in Figure 5.

7

A particularly strong fluorescence enhancement is exhibited by 3 when it is spotted on glass fiber paper[2,3] together with methyl benzilate. For the samples used to obtain the spectra in Figure 5, a solution of methyl benzilate in tetrahydrofuran was also

118

added to the polymerization mixture at the chain extension step. The presence of 7 increases the fluorescence intensity of the polymer containing 3 by about 80% and increases that of the polymer containing 6 by over 200%. The mechanism of fluorescence enhancement, perhaps by charge-transfer complex formation, is incompletely understood. Since methyl benzilate possesses a tertiary hydroxy group, it is possible that some reaction with isocyanate has taken place. It would be competing with 1,4-butanediol, a primary alcohol. It has been shown that primary alcohols can be 200 times more reactive with isocyanates than tertiary alcohols at 30°C in a catalyzed reaction[14].

CONCLUSIONS

Two fluorescent compounds, 2-diphenylacetyl-1,3-indanedione-1-(p-dimethylaminobenzaldazine), 3 and the bis(hydroxyethyl) analog 6 have been added to a transparent, aliphatic, amorphous polyurethane during the polymerization reaction. The fluorescence of 3 and 6 increase with polyurethane cure, a slightly greater effect being observed with 6, which reacts to become part of the polymer main chain.

The fluorescence of the dyes in the polyurethanes changes with temperature and can be used to identify thermal transitions in the polymers. When methyl benzilate is added during polymerization, a strong fluorescence enhancement is observed, the greater

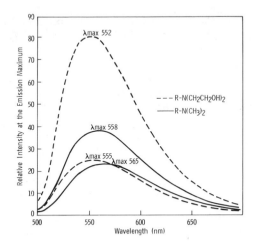

Figure 5. Fluorescence spectra of polyurethane sheets. Lower curves: 3.67×10^{-6} moles 3 or 6/100 g. Upper curves: 3.67 moles 3 or 6 and 2.06×10^{-5} moles methyl benzilate/100 g. The sheets were placed in a quartz cuvette and the Fluorolog was used in the right angle mode, with excitation at 476 nm.

effect occurring when 6 is the dye used. It has been demonstrated that a fluorescence enhancement like that observed on glass fiber paper can also occur in a polymer sheet. Many other organic compounds in addition to 7 could be used and diffusion experiments with enhancing compounds like 7 in solution should be done.

ACKNOWLEDGEMENTS

The authors would like to thank Drs. J. Taylor, M. A. Winnik and C. D. Eisenbach for helpful discussions concerning fluorescence spectroscopy in general and this research in particular.

REFERENCES

1. H. Morawetz, Science 203, 405 (1979).
2. E. J. Poziomek, E. V. Crabtree and J. W. Mullin, Anal. Lett. 14 (A11), 825 (1981).
3. E. J. Poziomek, E. V. Crabtree and R. A. MacKay, Anal. Lett 13 (A14), 1249 (1980).
4. R. A. Braun and W. A. Mosher, J. Am. Chem. Soc., 80, 3048 (1958).
5. A. K. Sen and C. C. Price, J. Org. Chem., 24, 125 (1959).
6. C. A. Byrne, E. J. Poziomek, O. I. Kutai, S. L. Suib and S. J. Huang, Polymer Preprints 24 (1), 79 (1983).
7. O. I. Kutai, C. A. Byrne, E. J. Poziomek, S. L. Suib and S. J. Huang, Polymer Materials Science and Engineering Proceedings, 49, 254 (1983).
8. C. A. Byrne, E. A. McHugh, R. W. Matton and N. S. Schneider, Org. Coat. Appl. Polym. Sci., 47, 49 (1982).
9. J. W. C. Van Bogart, D. A. Bluemke and S. L. Cooper, Polymer 22, 1428 (1981).
10. C. A. Byrne, D. P. Mack and M. A. Cleaves, Polymer Materials Science and Engineering Proceedings, 49, 58 (1983).
11. R. O. Loutfy, Macromolecules 14, 270 (1981).
12. R. L. Levy and D. P. Ames, Organic Coatings and Applied Polymer Science Proceedings 48, 116 (1983).
13. R. O. Loutfy and D. M. Teegarden, Macromolecules 16, 452 (1983).
14. J. H. Saunders and K. C. Frisch, "Polyurethanes: Chemistry and Technology", Part 1; Interscience, New York; 1962, p. 140.

AN XPS AND SEM STUDY OF POLYURETHANE SURFACES:

EXPERIMENTAL CONSIDERATIONS

R.W. Paynter* , B.D. Ratner* and H.R. Thomas[#]

Department of Chemical Engineering
   and Center for Bioengineering
University of Washington
Seattle, Washington 98195*

Pfizer Corporation, Easton, Pennsylvania 18042[#]

INTRODUCTION

Polyetherurethanes (PEU's) are possibly the only class of polymeric material with mechanical properties which make them suitable for long-term application in an implanted total artificial heart. Evidence suggests that the surface composition of PEU's influences, in a predictable manner, their interaction with blood (1,2). Both attenuated total reflection infrared (ATR-IR) (3,4) and X-ray photoelectron spectroscopy (XPS) (2,5,6,7,8) have shown that the surface compositions of these materials can differ considerably from the bulk, a surface excess of polyether segment being indicated. Since relationships between surface structure and blood compatibility have been established, an understanding of the factors which influence the nature of PEU surfaces might be used to fabricate PEU's with improved blood compatibility. In this paper, a new method for calculating and presenting quantitative depth profiles of atomic composition derived from angular dependent XPS studies is considered. The method is based upon fitting angular XPS profiles, calculated employing a novel algorithm, to experimental data. Results have been obtained for a series of PEU compositions, with casting solvent and extraction as additional variables.

In order to obtain meaningful XPS data, smooth surfaces must be utilized. This paper will consider solvent effects in the preparation of polyurethane cast films and will discuss the significance of the XPS results based upon SEM images of the cast films.

METHODS

The polyurethanes used in these experiments were either
synthesized in our laboratories or purchased. The synthesized
materials were formed by "capping" the terminal hydroxyl groups of
775 MW poly(propylene glycol) (PPG) with methylene diphenyl-
diisocyanate (MDI) and chain extending in solution the resulting
compound using ethylene diamine (9). The commercial material used
in these experiments was the poly(ether urea urethane), Biomer
(Ethicon), dried under vacuum to remove the dimethyl acetamide
(DMAc) in which it is shipped, redissolved in various high purity
solvents (Aldrich Gold Label, Burdick and Jackson "distilled in
glass", or Pierce) and filtered through Teflon microporous
filters.

Films were formed by centrifugally casting the polyurethane
solutions using a photoresist coater manufactured by Headway
Research, Garland, Texas. Small volumes of polymer solutions were
measured onto cleaned glass disks and spun at 4000 rpm for 20
seconds. Cast films were air-dried in a dust-free laminar flow
environment.

Scanning electron micrographs (SEM's) were taken on a JEOL
JSM-25 microscope. Specimens were sputter-coated with gold-
palladium prior to observation.

The surface structures of cast PEU films were modelled using
data obtained from angular dependent XPS studies. All spectra
were taken on a Hewlett Packard 5950B XPS instrument using an
aluminum monochromatized source at 800 W and a low-energy electron
floodgun to reduce charging. The sample was tilted in the XPS
instrument with a Surface Science Laboratories variable angle
probe to change the angle between the normal to the surface and
the electron collection optics (the photoelectron take-off angle,
θ). At each θ setting, the apparent composition of the sample
(expressed as atomic % of each element, on a hydrogen-free basis)
was determined. The most likely PEU surface structure was then
obtained by mathematical modelling using an integral solution of
the Beer-Lambert equation (10) (see Figure 1). The angular
dependence of the XPS signal from each element in the model was
calculated and expressed as a percentage of the total signal. The
figures obtained were then compared to the experimental data and
adjustments made in the model profile until good agreement was
achieved. Information about the depth dependence of atomic
composition was obtained this way. Bonding information was also
elucidated based upon a consideration of the C1s spectra resolved
into subpeaks by fitting Gaussian peaks of predetermined peak
width and binding energy shift (6,11).

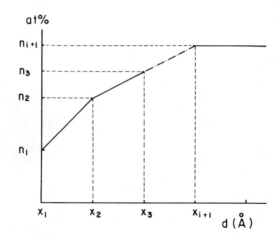

$$I(\theta)\alpha n_i\lambda\cos\theta + \lambda^2\cos^2\theta \sum_i \frac{(n_{i+1}-n_i)}{x_{i+1}-x_i}\left(e^{\frac{-x_i}{\lambda\cos\theta}} - e^{\frac{-x_{i+1}}{\lambda\cos\theta}}\right)$$

Figure 1. Scheme and algorithm for the calculation of the angular
dependence of the XPS signal (I) from an element whose
concentration (n) varies with depth (x) as specified.
λ = photoelectron inelastic mean free path. θ = sample
angle in the XPS instrument.

RESULTS AND DISCUSSION

SEM's of Biomer films cast from filtered solutions in
dimethyl formamide (DMF), DMAc, hexafluoroisopropanol (HFIP) and
o-cresol are shown in Figures 2, 3, 4 and 5, respectively. Film
quality is poor in DMF and DMAc. In HFIP, no detail in the
surface can be resolved even at 3000X magnification. In o-cresol,
only an occasional irregularity can be seen in generally smooth
cast films. This could indicate the relative solvent strengths
for Biomer of the solvents used. In poorer solvents, an increased
aggregation tendency (lower solubility) could explain the clumps
of material seen in the SEM's for DMF and DMAc cast films. It
should be noted that although these solvents were glass distilled
and used fresh from the bottle, the presence of water in them was
not systematically investigated. Water in these solvents would
certainly reduce their solvent power for PEU's. It is also of
interest that artificial hearts, blood pump diaphrams, and other
polyurethane prostheses have been fabricated by casting from DMF
and DMAc (9,12). Other casting solvents may improve the ability
to fabricate defect-free implant devices.

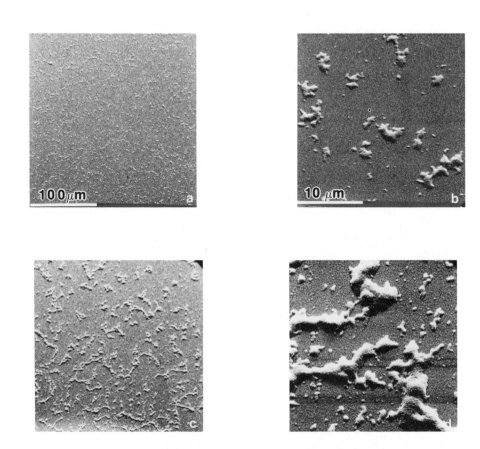

Figure 2. SEM micrographs of Biomer films centrifugally cast from DMF onto glass: (a) 0.5%, 300x; (b) 0.5%, 3000x; (c) 2%, 300x; (d) 2%, 3000x.

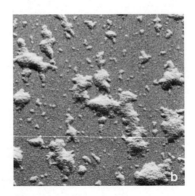

Figure 3. SEM micrographs of Miomer films centrifugally cast from DMAC onto glass: (a) 1%, 300x; (b) 1%, 3000x.

Figure 4. SEM micrographs of Biomer films centrifugally cast from HFIP onto glass: (a) 1%, 450x; (b) 1% 3000x.

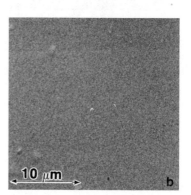

Figure 5. SEM micrographs of Biomer films centrifugally cast from o-cresol onto glass: (a) 1%, 450x; (b) 1%, 3000x.

Depth profile models derived from angular dependent XPS data for films of Biomer cast from 1% solutions in DMF, DMAc and HFIP onto glass are shown in Figures 6, 7 and 8, respectively. These depth profile diagrams are read in the following way. A vertical line placed at any position along the x-axis suggests the composition (in atomic %) at that depth into the polymer surface. The total length of the vertical line will be 100% composition. For each chemical component, the fraction of the vertical line subtended by the upper and lower boundaries for that chemical species is directly proportional to the model prediction of the atomic percent composition of that species. The sharp SiO_2 interface is readily observed in these models and the incorporation of such an interface in the model provides the best fit between the experimentally obtained angular dependent XPS data and the calculated data set derived from the model. The hard segment (indicated by the N signal) is seen to concentrate adjacent to the glass-polymer interface in each case. The concentration of nitrogen appears to decrease in each case as the air-polymer interface (the surface) is approached. Enrichment of polyether-like material and/or hydrocarbon-like material at the air interface is also commonly observed. A cast film of the PEU prepared in our laboratory with PPG 775 on glass shows no Si XPS signal (Figure 9). This would suggest that the film is thicker than the penetration depth for Si photoelectrons. However, once again, nitrogen is seen to be present in lower quantities at the air interface than within the bulk of the film and polyether is enriched at the surface. This surface polyether enrichment has been suggested by other workers as well (3,5), and can be explained by a minimization of interfacial energy at the air-polymer interface induced by the surface localization of this relatively low surface energy polymer.

The film thicknesses suggested by the best fit models for the Biomer cast from DMAc and DMF are unrealistically low and, particularly for the DMF case, do not seem consistent with possible molecular models. These results must be tempered by three considerations. First, inelastic mean free path (IMFP) values used in the equation shown in Figure 1 were taken from the compilation of Seah (13). However, controversy in the literature concerning the IMFP values would indicate that they could be off by at least 100%. Thus, the shape of the concentration profiles in the models shown in Figures 6-9 may be correct, but the values on the x-axis could scale up (or down) depending upon which set of IMFP valves was chosen. Second, the SEM's indicated that the DMF and DMAc films were of questionable quality. From films of this appearance, one would tend to have little confidence in the ability of the algorithm (which assumes very smooth surfaces) to meaningfully generate models. Thus, agreement between the experimentally obtained XPS angular dependent data set and the calculated profile may be fortuitous. Finally, this model is

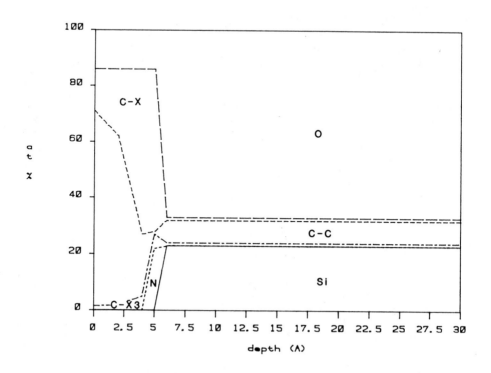

θ	EXPERIMENTAL						CALCULATED					
	C-C	C-X	C-X3	N	O	Si	C-C	C-X	C-X3	N	O	Si
30	15.2	8.3	1.4	1.0	55.3	18.8	15.1	8.4	1.5	0.8	55.4	18.9
62	20.5	15.1	2.1	1.2	45.7	15.4	20.4	13.2	1.8	1.2	47.5	15.9
80	35.7	23.2	2.3	1.4	28.8	8.7	35.7	23.1	2.3	1.7	29.0	8.3

Figure 6. Postulated depth profile obtained from Biomer centrifugally cast onto glass from a 1% solution in DMF. C-C
represents hydrocarbon-like functionalities resolved
from the C1s spectrum. C-X represents carbon species
singly bound to oxygen resolved from the C1s spectrum.
C-X3 represents carbon species bound with two or more
bonds to oxygen or nitrogen from the C1s spectrum.

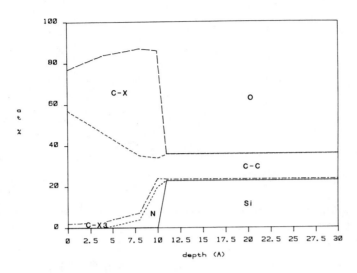

Figure 7. Postulated depth profile obtained for Biomer cast onto glass from a 1% solution in DMAc. See caption, Figure 6, for key.

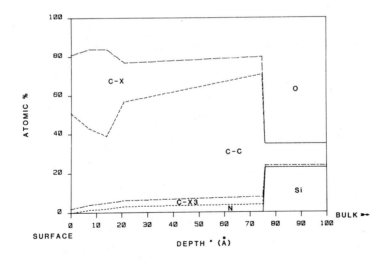

Figure 8. Postulated depth profile obtained for Biomer cast onto glass from a 1% solution in HFIP. See caption, Figure 6, for key.

129

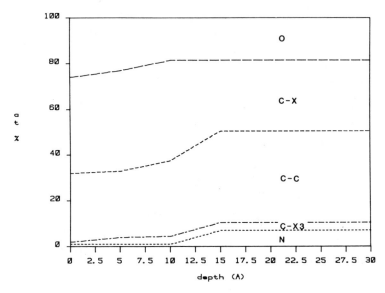

Figure 9. Postulated depth profile for unextracted PEU with 775 MW, PPG soft segment, cast onto glass from DMAc, air-facing side. See caption, Figure 6, for key.

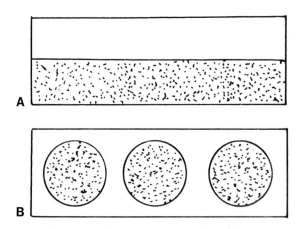

Figure 10. Schematic models for cross-sectioned surfaces containing equal fractions of hard and soft segment. The XPS angular dependent algorithm used in this work can treat Case A only.

designed to treat surfaces which vary in composition with depth, but do not vary in composition laterally (see Figure 10). The domains known to exist in PEU's or "aggregate" regions could significantly alter the ability of the models to accurately represent the physical reality of these materials. However, angular dependent XPS data can be applied to laterally inhomogeneous if appropriate models are developed (14).

CONCLUSIONS

The use of model depth profiles matched to angular dependent XPS data can help to enhance our understanding of PEU surface since more traditional XPS techniques for analysis of polymer surfaces average the concentrations of components over some 50-100 Å and do not give specific information about the outermost surface layer, the region at which biological interactions take place. However, such modeling techniques must always be used with a full awareness of the assumptions involved in the model. Careful SEM study of polymer films examined in the XPS experiment can help to set guidelines for how much significance should be attributed to the model. Other studies are ongoing in our laboratories to consider the physical significance of models suggested by this computational technique.

The depth profile models constructed from XPS data from smooth films indicates that there can be significant compositional differences between the outermost surface layers of a PEU and the polymer 30-75 Å beneath the surface. The implications of surface chemistry for blood and biological reaction with PEU's have already been discussed (1,2). Since these surface profiles should be strongly influenced by polymer composition, casting solvent, casting concentration, drying rate, interface composition, temperature, etc., and since few of these variable have been systematically explored, there is much research which remains to be performed before we will have a complete understanding of PEU surfaces. When this understanding has been accomplished, we will have the tools to systematically design PEU surfaces with enhanced biocompatibility.

ACKNOWLEDGEMENT

This work has been supported by NHLBI grant HL25951 and NIH DRR grant RR 01296.

REFERENCES

1. S.R. Hanson, L.A. Harker, B.D. Ratner and A.S. Hoffman, in: Biomaterials 1980, Advances in Biomaterials(G.D., Winter, D.F., Gibbons and H., Plenk, Jr, eds.), Vol. 3, pp.519-530, John Wiley and Sons, Chichester, England (1982).

2. V. Sa Da Costa, D. Brier-Russel, E.W. Salzman and E.W. Merrill, ESCA studies of polyurethanes: Blood platelet activation in relation to surface composition, J. Coll.Interf. Sci. 80, 445-452(1981).

3. K. Knutson and D.J. Lyman, in: Biomaterials: Interfacial Phenomena and Applications, ACS Advances in Chemistry Series(S.L., Cooper and N.A., Peppas, eds.), Vol. 199, pp. 109-132, American Chemical Society, Washingtion, D.C. (1982).

4. C.S.P. Sung and C.B. Hu, in: Multiphase Polymers, ACS Advances in Chemistry Series(S.L., Cooper and G.M., Estes, eds.), Vol. 176, pp.69-82, American Chemical Society, Washington, D.C. (1979).

5. C.S.P. Sung and C.B. Hu, ESCA studies of surface chemical composition of segmented polyurethanes, J. Biomed. Mater. Res. 13, 161-171(1979).

6. B.D. Ratner, in: Photon, Electron, and Ion Probes of Polymer Structure and Properties, ACS Symposium Series(D. W., Dwight, T.J., Fabish and H.R., Thomas, eds.), Vol. 162, pp.371-382, American Chemical Society, Washington, DC (1981).

7. S.W. Graham and D.M. Hercules, Surface spectroscopic studies of Biomer, J. Biomed. Mater. Res. 15, 465-477(1981).

8. M.D. Lelah, L.K. Lambrecht, B.R. Young and S.L. Cooper, Physiochemical characterization and in vivo blood tolerability of cast and extruded Biomer, J. Biomed. Mater. Res. 17, 1-22(1983).

9. J.L. Brash, B.K. Fritzinger and S.D. Bruck, Development of block copolyether-urethane intra-aortic balloons and other medical devices, J. Biomed. Mater. Res. 7, 313-334(1973).

10. R.W. Paynter, Modification of the Beer-Lambert equation for application to concentration gradients, Surf. Interf. Anal. 3, 186-187(1981).

11. D.T. Clark and H.R. Thomas, Applications of ESCA to polymer chemistry. X. Core and valence energy levels of a series of polyacrylates, J. Polym. Sci., Polym. Chem. Ed. 14, 1671-1700(1976).

12. W.M. Phillips, W.S. Pierce, G. Rosenberg and J.H. Donachy,
     in: Synthetic Biomedica;l Polymers(M., Szycher and W.J.,
     Robinson, eds.), pp.39-57, Technomic Publishing, Inc.,
     Westport, Conn. (1980).

13. M.P. Seah and W.A. Dench, Quantitative electron spectroscopy
     of surfaces: A standard data base for electron inelastic mean
     free paths in solids, Surf. Interf. Anal. 1, 2-11(1979).

14. H.R. Thomas and J.J. O'Malley, Surface studies on
     multicomponent polymer systems by X-ray photoelectron
     spectroscopy. Polystyrene/poly(ethylene oxide) diblock
     copolymers, Macromolecules 12, 323-329(1979).

POLYMER SURFACES POSSESSING MINIMAL INTERACTION
WITH BLOOD COMPONENTS

Y. Ikada, M. Suzuki, and Y. Tamada

Research Center for Medical Polymers and Biomaterials
Kyoto University
Kawara-cho, Shogoin, Sakyo-ku, Kyoto 606, Japan

INTRODUCTION

For molecular design of blood-compatible surfaces it is essential to have knowledge of the mechanism of clotting occurring on the foreign surfaces. The recent progress on the biochemistry of blood coagulation has disclosed much of primary reactions involved in the complicated cascade process of thrombus formation. Although platelet adhesion as well as the activation of Hageman factor (XII) by high-molecular-weight kininogen and prekallikrein, are not completely made clear in the molecular level, it is well known that the thrombus formation is triggered by an interaction between blood components and the foreign polymer surface[1]. It is not clear which blood component plays a decisive role in the trigger reaction, but the key reaction seems to involve some plasma proteins which will be adsorbed within a few seconds to a polymer surface brought into contact with blood. Platelet adhesion may follow the protein adsorption. Therefore, the adsorption of plasma proteins to polymer surfaces has been the objective of numerous investigations[2].

Our starting assumption in developing a blood-compatible material is that the polymer surface which interacts with blood to an insignificant extent may not induce the thrombus formation. In other words, the polymer surface which does not adsorb any plasma protein must be thromboresistant. It should be pointed out that the protein adsorption may not be a specific biological event (as an enzyme-substrate reaction) but more like a physico-chemical process. It is impossible to exaggerate that, among a large number of events occurring during the processes leading to thrombus formation, the physico-chemical adsorption is virtually

135

only one reaction that we can relatively readily regulate, unless any biologically active substances are used. Although there have been accumulated vast experimental results indicating that the protein adsorption depends greatly on the surface energy of substrates[3-6], little has been presented theoretically by correlating the surface energetics with protein adsorption. Therefore, we will attempt first to derive a theoretical expression for the work of adsorption in terms of the surface energy. Since the plasma proteins as well as the cells in blood interacting with the polymer surface are better regarded as two-dimensional, we use the term "work of adhesion" instead of work of adsorption.

In contrast to adhesion in air, only a few thermodynamic studies have been devoted to adhesion in biological systems. Recently van Oss[7] has pointed out that phagocytosis of bacteria by a phagocyte is independent of immune reactions but determined solely by the free energy changes during phagocytosis:

$$\Delta F_{net} = \gamma_{PB} - \gamma_{BW} \tag{1}$$

where γ_{BW} is the interfacial energy between a bacterium and water and γ_{PB} is the interfacial energy between the phagocyte and the bacterium. For instance, he and his coworkers[8] studied the phagocytosis of bacteria by platelets in aqueous media of different surface tensions and showed that bacterial ingestion should be increased with increasing bacterial surface tension if the liquid medium surface tension is lower than the surface tension of the platelets. According to Gerson and Scheer[9], the extent of adhesion of bacterial cells to hydrophobic plastic surfaces was increased as the difference in the free energy accompanying the cell attachment became larger (or more negative). The free energy change is given by

$$\Delta G_a = \gamma_{SM} + \gamma_{BM} - \gamma_{SB} \tag{2}$$

where γ_{SB}, γ_{SM}, and γ_{BM} are the interfacial free energies between the solid and the bacterium, between the solid and the aqueous medium, and between the bacterium and the aqueous medium, respectively. Such a thermodynamic approach to bacterial adhesion was also attempted by Dexter[10]. He assumed that through adsorption by bacteria, the original solid-water interface (SW) was replaced with a solid-adsorbed organic interface (SO) plus a diffuse organic-water interface (OW). Therefore, the change in the free energy is

$$\Delta F = \gamma_{SO} + \gamma_{OW} - \gamma_{SW} \tag{3}$$

In all the above equations it is very difficult to calculate the free energies, especially the interfacial energy between cells and the substrate in the aqueous environment.

136

WORK OF ADHESION IN AQUEOUS MEDIA

As is well known, the work of adhesion in vacuum or in air between body 1 and body 2 (W_{12}) is defined as

$$W_{12} = \gamma_1 + \gamma_2 - \gamma_{12} \tag{4}$$

where γ_1 and γ_2 are the surface free energies of body 1 and 2, respectively, and γ_{12} is the interfacial free energy between them. In analogy with W_{12}, the expression for the work of adhesion in water or in aqueous media ($W_{12,W}$) may be written as

$$W_{12,W} = \gamma_{1W} + \gamma_{2W} - [\gamma_{12}]_W \tag{5}$$

where γ_{1W} (γ_{2W}) is the free energy of the interface between body 1 (body 2) and water, and $[\gamma_{12}]_W$ is the interfacial energy of 1-2, the γ_{12} value in water. If both bodies have the surface free energy consisting of merely two components, that is, dispersive (γ^d) and polar (γ^p)

$$\gamma_1 = \gamma_1^d + \gamma_1^p \tag{6}$$

$$\gamma_2 = \gamma_2^d + \gamma_2^p \tag{7}$$

and W_{12} is tentatively assumed to be given by [11]

$$W_{12} = 2(\gamma_1^d \gamma_2^d)^{1/2} + 2(\gamma_1^p \gamma_2^p)^{1/2} \tag{8}$$

We may obtain the following equation for $W_{12,W}$, in analogy with eq.(8).

$$W_{12,W} = [2(\gamma_{1W}^d \gamma_{2W}^d)^{1/2} + 2(\gamma_{1W}^p \gamma_{2W}^p)^{1/2}]_W \tag{9}$$

Here the symbol $[\quad]_W$ denotes, as is in eq.(5), that the value in parenthesis is that not in vacuum but in a medium of water. It seems reasonable to express γ_{1W}^d as [12]

$$\gamma_{1W}^d = \gamma_1^d + \gamma_W^d - 2(\gamma_1^d \gamma_W^d)^{1/2} \tag{10}$$

This is rewritten as

$$\gamma_{1W}^d = \gamma_1^d [1 - \frac{2(\gamma_1^d \gamma_W^d)^{1/2}}{\gamma_1^d + \gamma_W^d}] + \gamma_W^d [1 - \frac{2(\gamma_1^d \gamma_W^d)^{1/2}}{\gamma_1^d + \gamma_W^d}] \tag{11}$$

The first term in the right-hand equation is the contribution from body 1 and the second term from water. On assuming that equations similar to eqs.(10) and (11) hold also for γ_{2W}^d, γ_{1W}^p, and γ_{2W}^p

and inserting these equations into eq.(9), we obtain

$$W_{12,W} = 2\{\gamma_1^d [1 - \frac{2(\gamma_1^d \gamma_W^d)^{1/2}}{\gamma_1^d + \gamma_W^d}]\}^{1/2} \{\gamma_2^d[1 - \frac{2(\gamma_2^d \gamma_W^d)^{1/2}}{\gamma_2^d + \gamma_W^d}]\}^{1/2}$$
$$+ 2\{\gamma_1^p [1 - \frac{2(\gamma_1^p \gamma_W^p)^{1/2}}{\gamma_1^p + \gamma_W^p}]\}^{1/2} \{\gamma_2^p[1 - \frac{2(\gamma_2^p \gamma_W^p)^{1/2}}{\gamma_2^p + \gamma_W^p}]\}^{1/2} \qquad (12)$$

In deriving the above equation, we inserted only the first terms of eq.(11) and the corresponding equations into eq.(9), since the second terms of these equations are related to water and may hardly contribute to $W_{12,W}$.

The work of adhesion would be as simple as

$$W'_{12,W} = \gamma_{1W} + \gamma_{2W} - \gamma_{12} \qquad (13)$$

provided that $[\gamma_{12}]_W = \gamma_{12}$. This is identical to eq.(2). In most cases, however, $[\gamma_{12}]_W$ is not equal to γ_{12}.

RESULTS

Calculation of $W_{12,W}$ from eq.(12) requires values of dispersive and polar components of γ_1, γ_2, and γ_W. Hereafter we designate body 1 to the polymer surface and body 2 to the adherent plasma protein. In Table 1 are listed γ_1 and γ_1^d which have been reported in the literature. γ_1^p can be calculated as $\gamma_1 - \gamma_1^d$. The γ_1 values for cellulose and poly(vinyl alcohol) (PVA) have not yet been determined and hence were calculated on assuming the geometric mean rule for γ_1^p. γ_{1W} is also given in Table 1. The values of γ_W used for the calculation are 21.8 and 51.0 erg·cm^{-2}, respectively[14]. The γ_{1W} value provides a quantitative measure for the "hydrophilicity" of the polymer; the material surface of higher γ_{1W} is more hydrophobic and that of $\gamma_{1W} = 0$ is completely hydrophilic.

For calculating $W_{12,W}$ we need to know γ_2^d and γ_2^p. In this work we tentatively employ the values for bovine serum albumin(BSA) reported by Paul and Sharma, that is, 31.4 and 33.6 erg·cm^{-2} for γ_2^d and γ_2^p, respectively[20]. Absolutely correct values for γ_2^d and γ_2^p are not necessary, as we merely intend to investigate qualitative dependence of $W_{12,W}$ on γ_{1W}.

Figure 1 shows the plot of $W_{12,W}$ against γ_{1W}. As is seen, the data scatter but apparently give a curve having a maximum. Two closed circles in Figure 1 refer to imaginary materials, "water" and "air". For both cases, $W_{12,W}$ is zero, γ_{1W} being 0 for water and 72.8 erg·cm^{-2} for air. Often, proteins are

138

Table 1. Surface free energies and work of adhesion with BSA in water for different polymers (unit in $erg \cdot cm^{-2}$)

	γ_1	γ_1^d	γ_1^p	γ_{1W}	$W_{12,W}$	Ref.
Polyethylene(HD)	35.0	35.0	0	52.6	1.40	13
Polyethylene(HD)	34.7	34.7	0	52.5	1.37	14
Polyethylene(LD)	33.2	33.2	0	52.2	1.20	15
Polypropylene	28.5	28.5	0	51.4	0.72	15
Polytetrafluoro-ethylene	19.5	19.5	0	51.2	0.25	13
Paraffin	25.5	25.5	0	51.1	0.40	13
Polytrifluoro-ethylene	23.4	23.4	0	51.0	0.17	14
Polystyrene	42.0	41.4	0.6	43.7	3.21	15
Polystyrene	44.8	43.6	1.2	40.3	3.82	14
Nylon 11	44.0	43.1	0.9	41.9	3.59	16
Nylon 11	43.8	42.8	1.0	41.2	3.63	14
Polyvinyl chloride	41.5	40.0	1.5	37.7	3.60	15
Polyethylene terephthalate	43.9	41.3	2.6	33.7	4.10	14
Polyethylene terephthalate	47.3	43.2	4.1	29.8	4.59	15
Polyvinylidene chloride	45.0	42.0	3.0	32.5	4.26	15
Polymethyl methacrylate	40.2	35.9	4.3	27.4	3.88	15
Nylon 6,6	47.0	40.8	6.2	24.6	4.58	15
Polyvinyl fluoride	36.7	31.3	5.4	24.1	3.54	15
Polyurethane-E	69.9	47.6	22.3	10.8	4.91	17
Polyurethane-T	63.4	43.7	19.7	11.1	4.69	17
Polyurethane-S	52.0	29.3	22.7	6.20	3.04	17
Cellulose	59	30	29	3.7	2.7	18
Polyvinyl alcohol	59.5	29	30.5	3.1	2.5	18
Canine artery	70.9	20.9	50.0	0.02	0.18	19
Canine vein	70.6	21.4	49.2	0.02	0.19	19
(Water)	72.8	21.8	51.0	0	0	
(Air)	0	0	0	72.8	0	

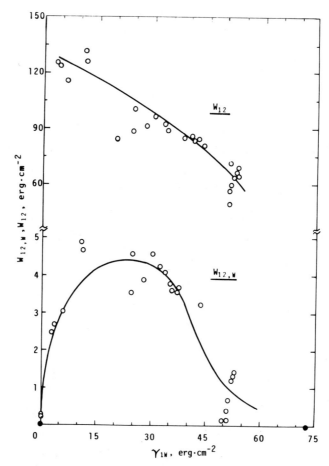

Figure 1. Work of adhesion with BSA in water ($W_{12,W}$) and in vacuum (W_{12}) as a function of interfacial free energy between water and polymer 1 (γ_{1W}).

positively adsorbed to an air/solution interface. Our calculation has nothing to do with this selective adsorption and predicts only that the work is zero for adhesion between air and protein. The result of $W_{12,W} = 0$ at $\gamma_{1W} = 0$ does not mean that the adhesion work between water and protein is zero, but simply that no adhesion would take place between the material "water" and the hydrated proteins in aqueous media. For comparison we calculated W_{12} according to eq.(4) and plotted the result in Figure 1. As is anticipated, the work of adhesion in vacuum

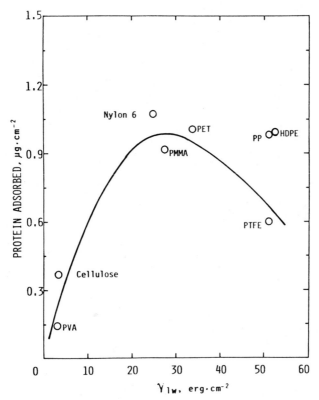

Figure 2. BSA adsorption to different surfaces at 37 °C in
phosphate buffered solution. The initial BSA
concentration is 3 mg·ml^{-1}. Abbreviation: PVA
(polyvinyl alcohol), PMMA(polymethyl methacrylate), PET(polyethylene terephthalate), PP(polypropylene), HDPE(high-density polyethylene),
PTFE(polytetrafluoroethylene)

is increased continuously as the hydrophilicity of the polymer
becomes higher. By contrast, the work of adhesion in water
derived from eq.(13) under the assumption of $[\gamma_{12}]_W = \gamma_{12}$ was
increased with the increasing hydrophobicity. This trend is
contrary to the general observation to indicate that eq.(13)
is not a correct expression for the work of adhesion in water.

For comparison with this theoretical prediction, we studied
BSA adsorption to various surfaces. They are PVA, cellulose,
nylon 6, poly(methyl methacrylate)(PMMA), poly(ethylene terephtha-

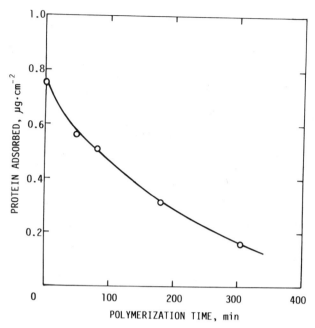

Figure 3. Adsorption of fibronectin onto polyethylene grafted
with acrylamide following argon-plasma treatment.
Adsorption was performed in 0.05 mg·ml^{-1} protein
solution.

late)(PET), polypropylene(PP), polyethylene(PE),and polytetraflu-
oroethylene(PTFE). The amount of protein adsorbed was determined
with the use of ^{131}I-labeled protein. The result is shown in
Figure 2. The PVA film has the minimal amount of protein adsorbed,
in accordance with our earlier result for a non-heat-treated
PVA[21]. Appreciable deviation of data for PP and PE might be
explained in terms of surface oxidation which readily occurs
on the polyolefins. The $W_{12,W}$ in Figure 1 for the materials
of high γ_{1W} such as about 50 erg·cm^{-2} seems to be somewhat too
small when we recall that proteins and cells attach to silicone
and PTFE to a comparatively significant extent[22]. The major
reason for this small $W_{12,W}$ may be due to unadequate selection
of the numerical value for γ_2^d and γ_2^p. Though the shape of
curve in Figure 1 is strongly dependent on the γ_2^d and γ_2^p values,
the maximum always appears in the $W_{12,W}-\gamma_{1W}$ plot, irrespective
of the γ_2 value adopted.

Figure 3 shows the influence of graft copolymerization

on fibronectin adsorption when the PE surface is graft-copolymerized with acrylamide(AAm). The graft copolymerization onto
the PE surface was carried out by the glow discharge technique[23].
It can be seen that protein adsorption diminishes as the graft
copolymerization is allowed to proceed.

DISCUSSION

There is a controversy on expressing the work of adhesion
in vacuum by eq.(8)[24]. γ_1^p is usually calculated by the use
of eq.(8) without any strong theoretical support, as·there
are few methods available for estimating γ_1^p at present. Therefore,
it should be kept in mind that eq.(12), derived under many
assumptions, is a very rough approximation for the work of
adhesion in water and, in addition, that the curve of Figure
1 is drawn using unsatisfactory experimental results. Nevertheless,
this curve seems to represent the features which have often
been observed. For instance, adhesion to highly hydrophilic
polymers with very low γ_{1W} takes place with difficulty in water,
whereas strong adhesion is attained for polar polymers with
medium γ_{1W} like dishes for cell culture[25]. We have no experimental
result for the case of very high γ_{1W} because of unavailability
of such highly hydrophobic polymers. This theoretical prediction
is apparently supported also by the experimental result given
in Figure 2. It follows that, at least qualitatively, the
adhesion phenomena in aqueous media can be discussed on the
basis of the $W_{12,W}-\gamma_{1W}$ plot in Figure 1, provided that only
dispersive and polar forces are operative in adhesion while
the ionic interaction is insignificant. It should be pointed
out that, if only two hydrophilic and hydrophobic surfaces
representing two extremes in surface energy are employed for
an adsorption or adhesion experiment, an erroneous conclusion
will be obtained because of nonlinear dependence of $W_{12,W}$ on γ_{1W}.

Figure 1 clearly demonstrates that there are two possibilities
for a polymer surface to have $W_{12,W}$ of zero, in other words,
to be nonadhesive. One is to create a superhydrophilic, in
other words, water-like surface ($\gamma_{1W}=0$) and the other, a superhydrophobic one ($\gamma_{1W}=73$ erg·cm^{-2}). This is an answer to the
old and still debated question why two extreme surfaces totally
different each other are relatively blood-compatible.

Of these two possibilities the hydrophilic surface appears
to be a more promising candidate for a long-term blood-compatible
polymer. The main reason is that it is almost impossible
to synthesize the hydrophobic polymer which exhibits higher
γ_{1W} than perfluoropolymers possessing the lowest γ_1 among the
conventional polymers. On the contrary, it may be much easier
to modify the polymer surface so as to have a $W_{12,W}$ close to

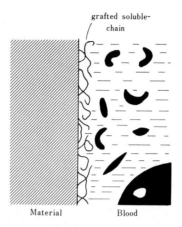

grafted soluble-
chain

Material   Blood

Figure 4. Schematic representation of the interface
between the diffuse surface and blood
components.

zero. For instance, surface grafting with water-soluble polymers
would yield such a material that has the water or aqueous layer
on the surface. The model structure is schematically shown
in Figure 4. We called this a diffuse surface[21]. Such a material
possesses a normal solid surface in the dried state, but produces
a diffuse interface when brought into contact with an aqueous
medium. The diffuse interface may contain a large amount of
water and a very small amount of water-soluble polymer chains
covalently bound to the substrate material of good mechanical
properties. It is likely that this two-phase system would
have γ_{1W} and $W_{12,W}$, both of which are very small or almost
close to zero, if the interface volume consists mostly of water.

Cell membranes having oligosaccharide chains on the surface
must have such a diffuse structure, but we do not intend to
mimic the surface structure of endothelium cells because they
presumably acquire the outstanding blood compatibility by virtue
of biologically active compounds such as prostacyclin[26]. Slippery
surfaces like mucous membranes may have also the diffuse structure
and give practically no physical stimulus to the surroundings.
Similarly, our diffuse surface will hardly stimulate the flowing
blood, in other words, will not significantly interact with
plasma proteins. Therefore, one can regard this as a nonstimulat-
ing bioinert surface. As is known, nonionic water-soluble
polymers such as poly(vinyl pyrrolidone), modified starch,
and dextran do not invoke thrombus formation when their aqueous
solutions are injected into the flowing blood and have been
used as plasma expander. Dilute solutions of modified PAAm

144

were also injected in pigeons to alter the blood flow[27]. Thus it is rather reasonable that the surface properly grafted with these water-soluble polymers may be blood-compatible. PAAm gels used for electrophoresis and dextran gels for gel permeation chromatography of proteins are known to adsorb an insignificant amount of proteins unless strongly crosslinked and ionized[28].

When the diffuse surface approaches another surface having similarly attached water-soluble chains, the interaction between the two surfaces will be not attractive but repulsive because of an increasingly repulsive osmotic and entropic interactions as the diffuse opposing polymer layers come into overlap in the good solvent[29]. This entropic repulsion due to hydrophilic polymers in the diffuse layer is also a large factor contributing to the minimal interaction with blood components for our super-hydrophilic surface. If the water content in the diffuse surface is decreased with such graft copolymerization as generating relatively densely grafted chains, γ_{1W} should become above zero and the entropic repulsion effect should be reduced.

It should be noted that the surface of the so-called hydrogels may have not a diffuse but rather clear surface even when brought into contact with water, if they contain water only by 50 to 80 wt%. As mentioned in the previous work,[21] the non-heat-treated PVA film of low crystallinity probably has a diffuse structure, although the overall water content is very low. The fact that the zeta potential of PVA remains practically null over a wide pH range supports the presence of a diffuse interface[21]. When AAm was graft-copolymerized onto PE, the zeta potential changed drastically from a large minus value to almost null[21] and excellent blood compatibility was found[30].

Detailed studies on the physical structure of the diffuse layer(*e.g.*, the length, the density of polymer chains...) and *in vivo* evaluation of the blood compatibility will be reported in the near future. Then we will be able to explain much more clearly why some research groups have found the graft copolymerization to be effective in improving the blood compatibility of biomaterials[31,32], while others have reported unusual biological interactions at radiation graft copolymer interfaces[33].

ABSTRACT

We have derived an expression for the work of adhesion of polymer material 1 with adherent 2 in a medium of water ($W_{12,W}$). When the $W_{12,W}$ values calculated for many conventional polymers from the derived equation are plotted against the interfacial free energy between material 1 and water (γ_{1W}), we obtain a curve having a maximum, indicating $W_{12,W}$ to approach zero if γ_{1W} is either zero or equal to γ_W. This suggests that the surface, either superhydrophilic ($\gamma_{1W}=0$) or superhydrophobic ($\gamma_{1W}=\gamma_W$) may possess excellent blood compatibility. It is almost impossible at present to develop such a material that has higher γ_{1W} than perfluoro polymers, whereas the surface having γ_{1W} close to zero may be more readily prepared by binding a water-soluble polymer layer on a material surface. For instance, protein adsorption onto a polyethylene film is decreased when acrylamide is graft-copolymerized onto the film surface. Existence of the diffuse interface between the material and the water medium can be evidenced by zeta potentials of the grafted substrate which are almost zero over a wide pH range, similar to a poly-(vinyl alcohol) film which exhibits insignificant protein adsorption.

REFERENCES

1. For instance, A.S. Hoffman, in "Biomaterials: Interfacial Phenomena and Applications", S.L. Cooper and N.A. Peppas, Eds., *ACS Adv. Chem.* **199**, 1 (1982).
2. For instance, "Interaction of the Blood with Natural and Artificial Surfaces", E.W. Salzman, Ed., Marcel Dekker, Inc. New York and Basel, 1981.
3. R.E. Bair, *Bull. N.Y. Acad. Med.*, **48**, 257 (1972).
4. B.D. Ratner, A.S. Hoffman, S.R. Hanson, L.A. Harker, and J.D. Whiffen, *J. Polym. Sci. Polym. Sym.* **66**, 363 (1979).
5. D.L. Coleman, D.E. Gregonis, and J.A. Andrade, *J. Biomed. Mater. Res.*, **16**, 381 (1982).
6. R.D. Bagnall and P.A. Arundel, *J. Biomed. Mater. Res.*, **17**, 459 (1983).
7. C.J. van Oss, *Ann. Rev. Microbiol.*, **32**, 19 (1978).
8. D.R. Absolom, D.W. Francis, W. Zingg, C.J. van Oss, and A.W. Neumann, *J. Coll. Interface Sci.*, **85**, 168 (1982).
9. D.F. Gerson and D. Scheer, *Biochem. Biophys. Acta*, **602**, 506 (1980).
10. S.C. Dexter, *J. Coll. Interface Sci.*, **70**, 346 (1979).
11. B.W. Cherry, *"Polymer Surfaces"*, Cambridge Univ. Press, 1981.
12. F.M. Fowkes, *Ind. Eng. Chem.*, **56**, 40 (1964).
13. F.M. Fowkes, *J. Phys. Chem.* **66**, 382 (1962); **67**, 2538 (1963); *ACS Adv. Chem.* **43**, 99 (1964); *Ind. Eng. Chem.* **56**, No.12, 40 (1964).
14. Y. Ikada and T. Matsunaga, *J. Adhesion Soc. Japan*, **15**, 91 (1979).
15. D.K. Owens and R.C. Wendt, *J. Appl. Polym. Sci.*, **13**, 1741 (1969).
16. R.H. Andrews and A.J. Kinloch, *Proc. Roy. Soc. London* A **332**, 385 (1973).
17. M.D. Lelah, R.J. Stafford, L.K. Lambrecht, B.R. Young, and S.L. Cooper, *Trans. Am. Soc. Artif. Intern. Organs* **27**, 504 (1981).

18. T. Matsunaga and Y. Ikada, *J. Coll. Interface Sci.*, <u>84</u>, 8 (1981).

19. U. Yano, T. Komai, T. Kawasaki, and Y. Hujiwara, *Prep. Japan Soc. Biomater.*, <u>2</u>, 45 (1980).

20. L. Paul and C.P. Sharma, *J. Coll. Interface Sci.*, <u>84</u>, 546 (1981).

21. Y. Ikada, H. Iwata, F. Horii, T. Matsunaga, M. Taniguchi, M. Suzuki, W. Taki, S. Yamagata, Y. Yonekawa, and H. Handa, *J. Biomed. Mater. Res.*, <u>15</u>, 697 (1981).

22. G.A. Bornzin and I.F. Miller, *J. Coll. Interface Sci.*, <u>86</u>, 539 (1982).

23. M. Suzuki, P. Dong-Skwi, Y. Ikada, *Polym. Preprints, Japan*, <u>31</u>, 286 (1982).

24. F.M. Fowkes and M.A. Mostafa, *Ind. Eng. Chem. Prod. Res. Dev.*, <u>17</u>, 3 (1978).

25. F. Grinnell, *Intern. Rev. Cytology*, <u>53</u>, 65 (1979).

26. S. Moncada, R. Gryglewski, S. Bunting, and J.R. Vane, *Nature* <u>263</u>, 663 (1976).

27. H.L. Greene, R.F. Mostardi, and R.F. Nokes, *Polym. Eng. Sci.*, <u>20</u>, 499 (1980).

28. J. Porath, *Biochem. Soc. Trans.*, <u>7</u>, 1197 (1979).

29. J. Klein and P. Luckham, *Nature*, <u>300</u>, 429 (1982).

30. Y. Ikada, M. Suzuki, M. Taniguchi, H. Iwata, W. Taki, H. Miyake, Y. Yonekawa, and H. Handa, *Radiat. Phys. Chem.*, <u>18</u>, 1207 (1981).

31. M. Suzuki, Y. Tamada, H. Iwata, and Y. Ikada, in *"Physicochemical Aspects of Polymer Surfaces"*, Vol.2, K.L. Mittal ed., Plenum Publ., 1983, p.923.

32. P.L. Kronick, in *"Synthetic Biomedical Polymers"*, M. Szycher and W.J. Robinson Eds., Technomic Publ., 1980, p.153.

33. A.S. Hoffman, B.D. Ratner, T.A. Horbett, L.O. Reynolds, and C. Su Cho, *Artificial Organs*, <u>7</u>(A), 86 (1983).

THERMODYNAMIC ASSESSMENT OF PLATELET ADHESION TO POLYACRYLAMIDE GELS

D.R. Absolom, M.H. Foo, W. Zingg and A.W. Neumann

Research Institute, The Hospital for Sick Children, Institute of Biomedical Engineering and Department of Mechanical Engineering, University of Toronto, Toronto, Ontario

INTRODUCTION

Hydrogels elicit a low cellular adhesion _in vitro_ when measured by platelet adhesion tests under static conditions (1,2). A thermodynamic model has been described which predicts the level of cellular adhesion and protein adsorption to a range of polymer surfaces (3,4). The aim of this model is to elucidate the possible mechanisms of cell adhesion and protein adsorption, and to use this knowledge for the development of non-thrombogenic biocompatible implant materials.

In a cell adhesion/protein adsorption study, the proper thermodynamic potential for the process of cell attachment is the grand canonical potential (5), which we call, for short, the free energy (6). In principle, bringing platelets into close proximity with the surface of the biomaterial is governed by physical laws of diffusion and turbulence. The subsequent attachment of platelets to the substrate can be described by thermodynamic means. A hypothetical process, for instance cellular attachment, will be favoured if the net change in the free energy for that process is negative. On the other hand, such a hypothetical process is unlikely to occur if the net change in free energy is positive. The creation or annihilation of the various interfaces involved in the process of cellular adhesion will be reflected in a change in the free energy for that process, in our case platelet adhesion. Therefore, the interfacial tensions will appear in the expressions for such a change in free energy as shown in the equation describing the energy balance for the process of cell adhesion:

$$\Delta F^{adh} = \gamma_{PS} - \gamma_{PL} - \gamma_{SL} \qquad [1]$$
$$= f(\gamma_{PV}, \gamma_{SV}) - f(\gamma_{PV}, \gamma_{LV}) - f(\gamma_{SV}, \gamma_{LV}) \qquad [2]$$

where γ_{PS} is the interfacial tension between platelet and surface, γ_{PL} is the interfacial tension between platelet and liquid suspension, and γ_{SL} is the interfacial tension between surface and liquid. Thus, in order to utilize the above thermodynamic predictions, it is necessary to determine the interfacial tensions γ_{PS}, γ_{PL} and γ_{SL}. While the interfacial tensions between liquid/vapor phases can be easily obtained by several methods, this is not so for interfacial tensions involving the solid phases. The most appropriate method for the determination of the surface tension of a polymer surface is through contact angle measurements.

The theoretical background and the methodology of contact angle measurements have been published previously (7). The determination of contact angles on very hydrophilic surfaces, such as hydrogels, introduces a number of specific difficulties. In this paper we describe a method for such measurements. Based on these measurements it is then possible to calculate the interfacial tensions which are relevant for platelet adhesion and thus to predict from free energy considerations the _relative_ extent of platelet adhesion to a series of hydrogels having different surface properties. Platelet adhesion to a series of polyacrylamide gels with different surface tensions was then determined experimentally. It was thus possible to compare the experimental results with the predicted changes in ΔF^{adh}. It was shown that platelet adhesion to hydrogels indeed follows the predictions of the thermodynamic model.

MATERIALS AND METHODS

1. Preparation of Test Surfaces: Polyacrylamide gels was prepared by polymerizing an aqueous solution of purified acrylamide monomer with the monomer cross-linking agent bis(N,N'-methylene bisacrylamide) using a free radical generating catalyst system following standard procedures (8). The catalysts were ammonium persulfate as the initiator and N,N,N',N',-tetramethylethylene diamide (TEMED) as a regulator. Fresh solutions of gel mixture and ammonium persulfate were prepared for each experiment. In order to produce gels with different surface properties the concentration of the total gel composition and percentage of cross-linking was varied.

The gel mixtures used in the preparation of surfaces were characterized by the concentration of the monomer and the percentage of cross-linking reagent in the monomer mixture. For this characterization we used the equations of Hjérten (9) to calculate the gel composition; %T denotes the total percentage of both monomers (acrylamide and bis-acrylamide) and %C is the percentage concentration of the cross-linker relative to the total concentration (cf. Legend to Fig. 1).

Before the polymerization reaction, the mixture was subjected to vacuum suction in order to degas the solution as the polymerization reaction is strongly retarded by oxygen. Then, the mixture was poured into a petri dish. After the gel formed, the gel was soaked in saline solution to which 0.02% sodium azide has been added as a bacteriostatic agent for 4 days, in order to remove completely the unreacted monomers in the gel and to ensure that the gel has obtained equilibrium with the saline solution. The contact angle and water content of the test sample were then determined.

2. Determination of the Surface Tension and Contact Angle: The surface tension of the suspending liquid (γ_{LV}) before and after platelet adhesion was measured by means of a Wilhelmy balance (10). Time dependent contact angles were determined at 10 minute intervals with 10 microlitre drops of saline (0.15 $\underline{M}$ NaCl) made with deionized distilled water, applied with a microsyringe. For the contact angle measurement a small telescope was used with built-in goniometer (Spindler and Hoyer, Göttingen, F.R.G.). A minimum of 20 readings are used to establish the values reported in this study. Five drops of saline were applied at approximately the same time to the gel surface and the advancing contact angle measured twice on both the left and right hand side of each drop. At each interval fresh drops of saline were applied to previously unoccupied sites of the gel surface.

In the case of platelets contact angle measurements were performed with drops of saline on layers of platelets deposited by suction on anisotropic cellulose acetate membranes as described previously (11).

3. Preparation of Washed Pig Platelet Suspension: The platelets from pigs were prepared according to the procedure of Mustard et al (12). After the final centrifugation the platelets were suspended in protein free Tyrode buffer and the platelet concentration determined on a Coulter Counter. The final platelet concentration was adjusted by dilution to give a cell suspension of 2×10^8 platelets/ml. This standardized cell concentration was used for all experiments. The viability of the platelets was assessed by means of adenosine diphosphate (ADP) induced aggregation using a Payton Dual Channel Aggregometer.

4. Static Platelet Adhesion Tests: 1.0 ml of the platelet suspension was then exposed to the various hydrogel surfaces for 4 minutes. The cell suspension was retained in wells machined in teflon moulds giving rise to an exposed surface area of 2.5 cm^2 Twenty-four of these areas were examined for each hydrogel surface giving rise to a total surface area examined of 60.0 cm^2 derived from six separate hydrogel preparations. For each preparation, the surface was examined in duplicate with three wells per specimen. During the incubation period, the platelet suspension was maintained at 37°C and no agitation was employed. Following incubation the teflon moulds were removed and the hydrogel surfaces rinsed without exposure to an air interface, by means of a dilution/displacement technique under constant shear conditions in a specially designed apparatus described in detail in a previous publication (13). Following rinsing, the number of adhering platelets per unit surface area of hydrogel preparation was determined by microscopic inspection of the surface using a fully automatic computerized digital image analysis system (Omnicon 3000, Bausch and Lomb, Rochester, N.Y.) The major advantage of this system is that the analysis is performed using automatic stage motion and as such all operator subjectivity and bias as to which microscopic fields etc., are analyzed is completely removed. The surface area analyzed is dictated by the automatic stage motion option which operates in a rectilinear two-dimensional pattern. Consequently, the surface area examined consisted of the largest square which could be circumscribed by the well diameter. This choice has the added advantage that it necessitated the exclusion of those areas in close proximity with the teflon walls, defining the well regions, and thereby obviates concerns due to the so-called "wall effects." Using this system 64% of the total surface area for each well was examined.

5. Water Content and Swelling Factor Measurement of Gel Samples: The various gel samples were blotted gently with filter paper to remove excess surface water. The gel samples were then weighed and air dried in an oven at 45°C for two days. The gel samples were then reweighed and the water content and extent of swelling were determined (cf. Legend to Fig. 6).

RESULTS

1. Surface Characterization of the Polyacrylamide Gels: A consistent pattern of advancing contact angle measurements was observed in the series of polyacrylamide co-polymers. The contact angle is plotted as a function of time. The curve can be presented in several regions. The first region, from time zero to the time when the plateau region is reached, occurs as the

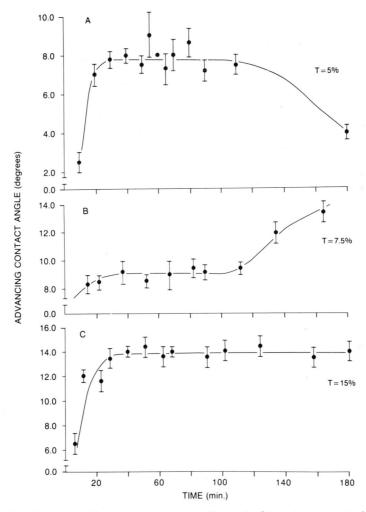

Figure 1   Contact angle measurements on polyacrylamide gels characterized by
the gel composition, as a function of time.
A. %T = 5; %C = 5; B. %T = 7.5; %C = 5; C. %T = 15; %C = 5

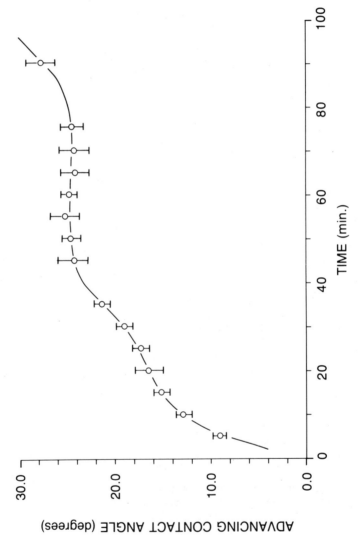

Figure 2   Contact angle measurements on a layer of pig platelets as a function of time.

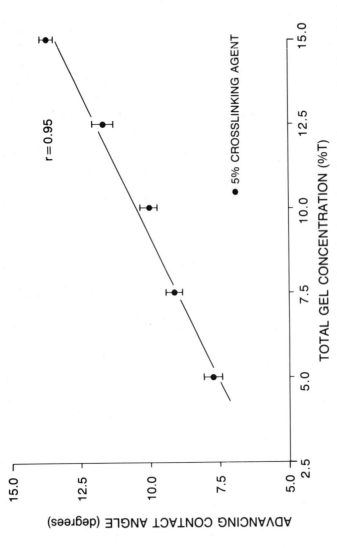

Figure 3   Plateau value of the advancing contact angle of polyacrylamide gels as a function of total gel composition (%T), with a fixed degree of crosslinking, %C = 5.

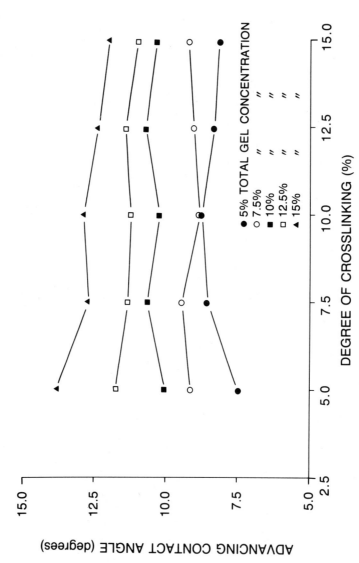

Figure 4 Plateau value of the advancing contact angle of polyacrylamide gels as a function of the degree of crosslinking (%C), with various total gel concentrations, %T.

advancing contact angle continually increases with time as a result of evaporation of excess water from the gel surface. In region II, the plateau region, the contact angle represents the contact angle which is characteristic for the surface of the substrate. The third region appears where the contact angle is seen to be either increasing or decreasing with time, due to physical and/or chemical changes in the gel. Some typical contact angle measurements on gels of different compositions are shown in Figure 1. Figure 2 shows the data for pig platelets. We take the plateau region to represent the effective contact angle (θ) value for the surface of interest. This plateau value is generally stable for 20–40 minutes and is used to determine the surface tension of the surface via the equation of state approach (14). It is the plateau contact angle value which is used to illustrate differences in the gel surfaces in Figures 3–5.

Figure 3 shows that the advancing contact angle increases linearly with increasing total gel concentration (%T) when the degree of cross-linking (%C) is held constant. In order to investigate this further, we wished to establish whether varying the degree of cross-linking affected the measured plateau value of the contact angle when the total gel concentration was held constant. Figure 4 depicts the results for a series of polyacrylamide gels where the extent of cross-linking was varied from 5% to 15% and in which %T was held constant. Within experimental limits no affect of varying %C on the advancing contact angle could be observed.

In Figure 5 the advancing contact angle measurements (θ) from the plateau region are plotted as a function of %T for a range of gel compositions in which both %T and %C have been varied. It is clear from this figure that whilst θ increases with increasing %T, that varying %C in the range from 5% to 15% had no significant effect on the plateau value of the contact angle for any of the various gel combinations investigated.

The next phase in this work was to determine whether any relationship existed between the hydrophobicity of the gel surface and the degree of swelling of the gel. A linear relation between the degree of swelling and hydrophilicity was seen as indicated in Figure 6.

2. Theoretical Prediction of ΔF^{adh}: The final phase in this work was to establish whether the surface tension data for the various interacting species (gels, platelets and suspending buffer) could be used to develop a theoretical prediction of the relative extent of platelet adhesion to the various hydrogel surfaces. For this purpose the contact angle measurements were

157

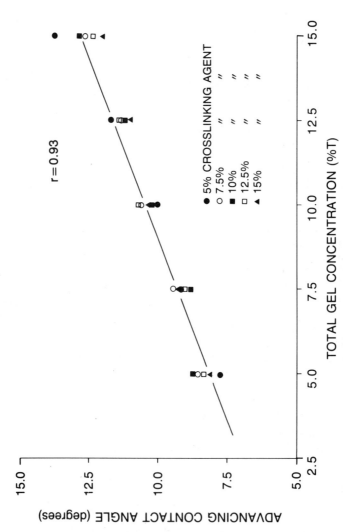

Figure 5   A summary of the plateau value of advancing contact angle measurements on a series of polyacrylamide gels as a function of total gel concentration (%T), with various degrees of crosslinking (%C).

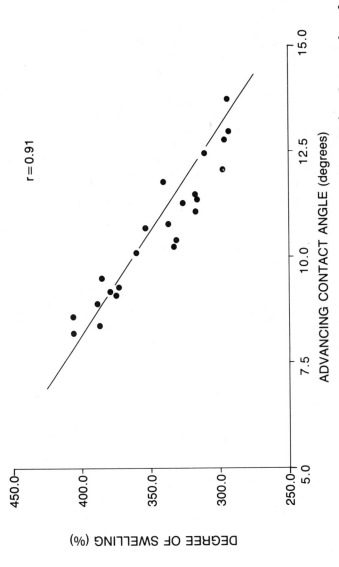

Figure 6   Swelling degree of polyacrylamide gels as a function of the corresponding plateau value of the advancing contact angle.

159

used to obtain the surface tension of the polyacrylamide gel surfaces, γ_{SV}, and of the platelets, γ_{PV}, via an equation of state approach described in detail elsewhere (14,15). Using this approach a contact angle of $\theta = 24.8°$ for platelets yields a surface tension for the platelets, $\gamma_{PV} = 66.7$ ergs cm^2. Similarly the various polyacrylamides have experimentally determined surface tensions γ_{SV}, ranging from 70.0 to 72.2 ergs/cm^2. This information in conjunction with the surface tension of the suspending buffer, $\gamma_{LV} = 72.8$ ergs/cm^2, enables us to calculate the net change in free energy (ΔF^{adh}) for the process of platelet adhesion to the various gel surfaces. Such a theoretical plot based on the experimental data obtained during the course of this work is given in Figure 7. The calculations indicate that the net free energy change for the process of platelet adhesion (ΔF^{adh}) increases with increasing surface tension of the hydrogel surface. According to this prediction, the extent of platelet adhesion should decrease with increasing substrate surface tension, γ_{SV}, when the surface tension of the buffer, γ_{LV}, is greater than the surface tension of the adhering platelets, γ_{PV}, i.e. when $\gamma_{LV} > \gamma_{PV}$.

Finally, we wished to determine whether these theoretical predictions were confirmed by experimental observation. Plotted in Figure 8 is the experimentally determined platelet adhesion data as a function of the substrate surface tension. The experimental curve confirms the theoretical predictions indicating that the number of platelets adhering per unit surface area decreases in a linear manner with increasing substrate surface tension. Linear regression of the data yields an intercept with the γ_{SV} axis at $\gamma_{SV} = 72.9$ ergs/cm^2. The model implies that this value should be equal to the surface tension of the suspending buffer, which it in fact is. The correlation coefficient for the straight line fit of the platelet adhesion data is 0.92.

DISCUSSION

The thermodynamic model used in this work has been applied to various materials with a wide range of substrate surface tensions (15). Excellent agreement has been found between the theoretical predictions and the experimental data for granulocytes (3,16,17), erythrocytes (18) and various strains of bacteria (19). In this work we have extended this investigation to include hydrogels as the substrate surface. Hydrogels typically have a high surface tension of the order of 70 ergs/cm^2 or higher. A good correlation was found between the hydrophobicity and the total gel content (%T), regardless of the extent of cross-linking (%C) and between the surface hydrophobicity and the extent of gel swelling.

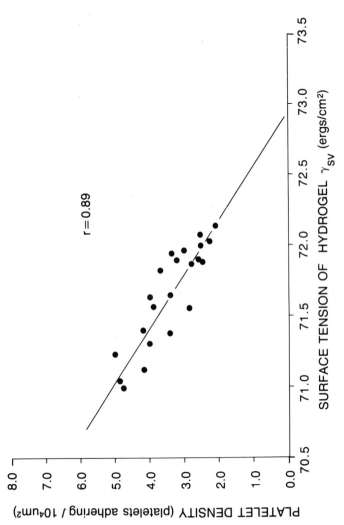

Figure 8   Platelet adhesion to various polyacrylamide gels as a function of substrate surface tension, γSV.

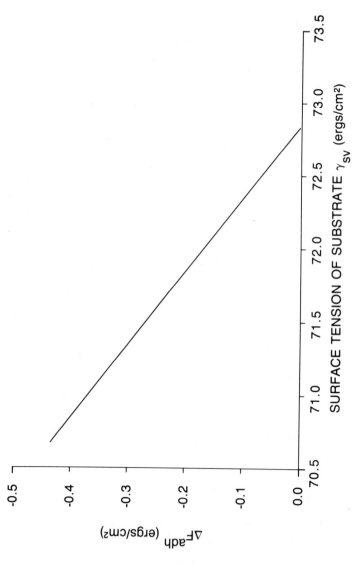

Figure 7   Free energy of adhesion versus the substrate surface tension, γ_{SV}. For this calculation, γ_{LV} = 72.8 ergs/cm^2 and γ_{PV} = 66.7 ergs/cm^2 were used.

Finally, it was experimentally determined that the extent of platelet adhesion increased linearly with increasing substrate surface hydrophobicity, i.e. the number of platelets adhering per unit surface area of gel decreased linearly with increasing substrate surface tension. This experimentally observed pattern of platelet adhesion is in accord with the thermodynamic predictions. The extent of platelet adhesion was also found to be dependent of the extent of gel swelling. Since a direct linear correlation was found between the surface tension of the gel and the extent of swelling, it is likely that the relationship between platelet adhesion and gel swelling is probably a manifestation of the surface properties of the gel.

It may be useful to reiterate that the extent of platelet adhesion per unit hydrogel surface area was determined after a fixed but nevertheless arbitarily chosen contact time of 4 minutes. Accordingly, the question might be raised as to whether our experimental platelet adhesion data reflects a rate of reaction which on some hydrogel surfaces was faster than on others thereby invalidating the present thermodynamic analysis which a priori does not provide kinetic information with respect to cell adhesion. In response, we should like to refer the interested reader to two previous publications from our group in which we have discussed in detail kinetic versus thermodynamic aspects of platelet adhesion to various polymer surfaces (20,21). In that work, we have shown that whilst kinetic affects determine the extent of platelet adhesion at arbitarily defined assay times, the overall relative affinity of platelets (as defined by adherent platelet density) for all of the various polymer surface examined was maintained for all of the contact times investigated. That is, the experimental data indicates that the overall extent of platelet adhesion is always larger on the higher energy surfaces than on the lower energy surfaces regardless of the contact time interval used. This suggests that the relative primary adhesive affinity of platelets for the high energy surfaces is always greater than the affinity of platelets for the low energy surfaces. This statement is supported by the fact that enhanced platelet adhesion is experimentally observed on the more hydrophilic polymer surfaces and as such lends considerable weight to the thermodynamic interpretation of platelet adhesion reactivity as outlined earlier in this article regardless of the arbitarily chosen incubation period.

The agreement between the directly measured γ_{LV} value of $\gamma_{LV} = 72.8$ ergs/cm^2 with γ_{SV}-axis intercept ($\gamma_{SV} = 72.9$ ergs/cm^2) suggests that any macromolecules which may be exuded by the platelets, if any, are not playing a role in determining the operative surface tension of the suspending fluid. Further pre-

163

liminary studies appear to confirm this contention. When serum proteins such as albumin, fibrinogen or γ-globulin G are incorporated in the buffer solution, the presence of these proteins in bulk solution do <u>not</u> appear to either enhance or reduce the extent of platelet adhesion to the various hydrogel surfaces, as compared to the protein free control experiments. However, when the gel surface is <u>pre</u>coated with the various proteins, considerable effect on the extent of platelet adhesion is observed. Fibrinogen and γ-globulin precoating greatly enhances the extent of platelet adhesion to the hydrogels whereas albumin slightly reduces platelet adhesion. It is not yet clear whether these effects are solely due to changes in γ_{SV} due to protein adsorption.

SUMMARY

The purpose of this study was to establish a procedure for the determination of contact angles and surface tensions of hydrogels and to test the applicability of a thermodynamic model to the adhesion of platelets to such gels. By controlling the gel concentration and the degree of swelling, hydrogels were produced with surface tensions ranging from 70 to 72.2 ergs/cm^2. The plateaux observed in the contact angle vs. time curves were identified as the relevant contact angles and proved to be highly consistent. Excellent quantitative agreement was obtained between thermodynamic model and experimentally observed extent of platelet adhesion.

REFERENCES

1. A.S. Hoffman, G. Schmer, C. Harris et al. Trans. Am. Soc. Artif. Intern. Organs <u>18</u>, 10, (1972).
2. S.D. Bruck. J. Biomed. Mat. Res. <u>7</u>, 387, (1973).
3. D.R. Absolom, A.W. Neumann, W. Zingg et al. Trans. Am. Soc. Artif. Intern. Organs <u>25</u>, 152, (1979).
4. A.W. Neumann, D.R. Absolom, W. Zingg, et al. Surface Thermodynamics of protein adsorption, cell adhesion and cell engulfment. in: Biocomptaible Polymers, M. Syzcher, ed. Thermo. Electron, Walton, Mass. Chapter 3, (1983).
5. A.W. Adamson. Physical Chemistry of Surfaces. pp 27–28, J.W. Wiley and Sons, N.Y., (1982).
6. A.W. Neumann, J.K. Spelt and D.R. Absolom. Measurement of the surface tensions of blood cells and other particles. Annals New York Acad. Sci. (In Press).
7. A.W. Neumann. Adv. Colloid Interface Sci. <u>4</u>, 105, (1974).
8. R.C. Allen and H.R. Mauer. in: <u>Electrophoresis and isoelectric Focussing in Polyacrylamide Gels</u>. W. de Gruyter, New York, 1974. pp 16–27.

9. S. Hjerten. Arch. Biochem. Biophys. 1, 147, (1962).
10. O. Driedger, A.W. Neumann, and P.J. Sell. Kolloid Z.Z. Polym. 201, 52, (1965).
11. D.R. Absolom, D.W. Francis, W. Zingg, et al. J. Colloid Interface Sci. 85, 168, (1982).
12. J.F. Mustard, D.W. Perry, M.G. Ardlie et al. Brit. J. Haematol. 22. 193, (1972).
13. O.S. Hum, A.W. Neumann and W. Zingg. Thromb. Res. 7, 461, (1975).
14. A.W. Neumann, R.J. Good, C.J. Hope et al. J. Colloid Interface Sci. 49, 291, (1974).
15. D.R. Absolom, W. Zingg, C.J. van Oss et al. Measurement of surface tension of cells and biopolymers with application to the evaluation of biocompatibility. in: Comprehensive Biotechnology, Vol III, Chapter 23, C.W. Robinson ed., Blackwell, Oxford, In Press.
16. A.W. Neumann, D.R. Absolom, W. Zingg et al. Cell Biophysics, 1, 79, (1979).
17. D.R. Absolom, C.J. van Oss, R.J. Genco et al. Cell Biophysics 2, 113, (1980).
18. D.R. Absolom, W. Zingg, C. Thompson et al. Erythrocyte Adhesion to Polymer Surfaces. I. J. Colloid Interface Sci. (In Press).
19. D.R. Absolom, F.V. Lamberti, Z. Policova et al. J. Appld. Environ. Microbiology 46, 90, (1983).
20. A.W. Neumann, O.S. Hum, D.W. Francis et al. J. Biomed. Mater. Res. 14, 499, (1980).
21. A.W. Neumann, D.W. Franciss, W. Zingg et al. J. Biomed. Mater. Res. 17, 375, (1983).
22. M.F. Refojo. in: Hydrogels for Medical and Related Applications; J.D. Andrade, ed. A.C.S. Symposium Series 31, A.C.S. Washington, (1976).

REPRODUCIBLE RESPONSE OF CERTAIN POLYMERS TO

CHANGES IN THE SURROUNDING ENVIRONMENT

J. Heller
Polymer Sciences Department
SRI International
Menlo Park, CA 94025

This chapter briefly reviews a number of polymer systems that undergo a significant and reproducible change in properties in response to relatively minor changes in the external environment. The chapter is not intended to be an exhaustive review, but rather focuses on polymer systems that may have potential applications in the field of self-regulated drug delivery devices. Two types of polymers are considered: polymers that undergo reversible dimensional changes and polymers that undergo changes in the rate of erosion.

Polymers that Undergo Dimensional Changes

Polymers that undergo reversible dimensional changes can have applications in self-regulated drug delivery devices if the therapeutic agent is contained in a core that is surrounded by a membrane constructed from a polymer capable of undergoing changes in permeability as a direct consequence of its interaction with the external environment.

One of the oldest and best known examples of a polymer that undergoes marked dimensional changes with changes in the external pH is poly(acrylic acid) partially esterified with poly(vinyl alcohol) (1-4). The films are cast from aqueous solutions containing poly(acrylic acid) and poly(vinyl alcohol) and are then heated to induce crosslinking by esterification between the carbonyl and hydroxyl groups. When strips are cut from the crosslinked films and placed in water, they undergo reversible changes in length of about 100% when exposed to acid and then to base. This effect is shown in Figure 1. The expansion and contraction of the polymer network has been ascribed to

electrostatic repulsion of charges fixed in the network or to
Donnan osmotic forces. A system based on a partially hydrolyzed
poly(methyl acrylate) has also been described (5).

Another interesting system is that prepared from crosslinked
poly(methacrylic acid) which is allowed to interact with a low
molecular weight poly(ethylene glycol) (6-13). In this polymer
system, a complexation through cooperative hydrogen bonding takes
place as follows:

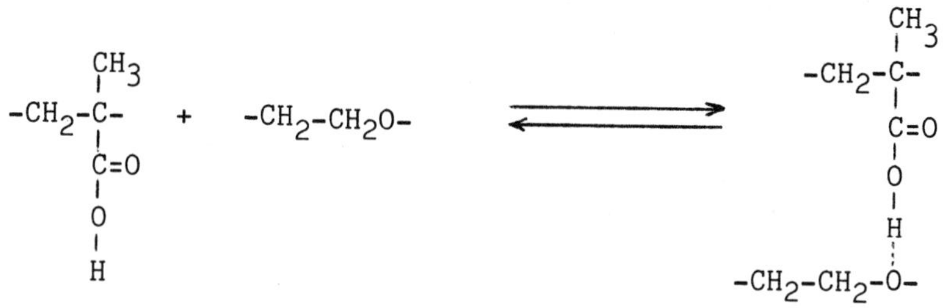

$$
\begin{array}{ccccc}
& CH_3 & & & CH_3 \\
& | & & & | \\
-CH_2-C- & + & -CH_2-CH_2O- & \rightleftharpoons & -CH_2-C- \\
& | & & & | \\
& C=O & & & C=O \\
& | & & & | \\
& O & & & O \\
& | & & & | \\
& H & & & H \\
& & & & \vdots \\
& & & & -CH_2-CH_2-O-
\end{array}
$$

When a crosslinked poly(methacrylic acid) membrane is placed
in water and a small amount of low molecular weight poly(ethylene
glycol) is added, a significant contraction of the membrane takes
place. If the dimensional changes of the poly(methacrylic acid)
membrane are mechanically constrained, addition of poly(ethylene
glycol) creates stresses within the membrane, and these can lead
to significant changes in permeability (13). Figure 2 shows a
dramatic increase in diffusion of albumin through a
poly(methacrylic acid) (PMAA) membrane on addition of small
amounts of poly(ethylene glycol) (PEG). When the membrane is
rinsed with a pH 8 water solution, the complex can be dissociated
and the membrane again becomes impermeable to albumin. Such a
system has been termed a chemical valve (13).

In addition to the hydrogen bonding, it has been postulated
that an endothermic hydrophobic interaction also takes place
between the methyl group in poly(methacrylic acid) and the
ethylene backbone of poly(ethylene glycol). An interesting
consequence of this hydrophobic interaction is shown in Figure 3,
which shows a reversible contraction of the membrane as a function
of temperature (9, 12). This contraction is particularly
pronounced in the region $20^{\circ}-30^{\circ}C$, where changes in length of 40%
occur with a $10^{\circ}C$ change in temperature.

Yet another interesting polymer system is one that undergoes
dimensional changes upon exposure to light (14). The dimensional
changes of the polymer are based on isomerization of
indolinospirobenzopyran shown below:

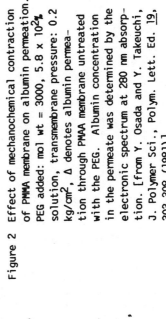

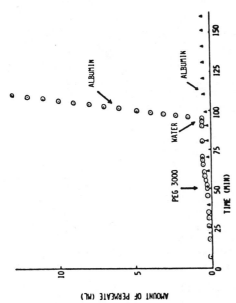

AMOUNT OF PERMEATE (ML)

Figure 2   Effect of mechanochemical contraction
of PMMA membrane on albumin permeation.
PEG added: mol wt = 3000, 5.8 x 10^2%
solution, transmembrane pressure: 0.2
kg/cm^2, Δ denotes albumin permeation through PMMA membrane untreated
with the PEG. Albumin concentration
in the permeate was determined by the
electronic spectrum at 280 nm absorption. [from Y. Osada and Y. Takeuchi,
J. Polymer Sci., Polym. Lett. Ed. 19,
303-308 (1981)]

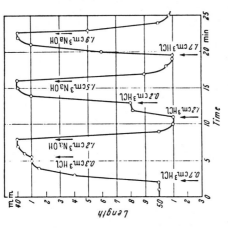

Figure 1   Change in length, with addition of acid or base, of
filaments prepared from 80% poly(vinyl alcohol) and 20%
poly(acrylic acid), made water-insoluble by stretch-
vulcanization. Ordinate: length of the filaments in
millimeters.   Abscissa: time in minutes.   The moments
at which acid or base are added are marked by arrows.
Dry weight of filaments: 6 mg.   Constant load on filaments: 360 mg.   Embedding fluid: 10 cm^3 of 0.01
n NaCl solution in water. Added solutions: 0.02 N HCl,
or 0.02 N NaOH. [from W. Kuhn, et al, Fortschr.
Hochpolym. - Forsch., 1, 540-592 (1960)].

169

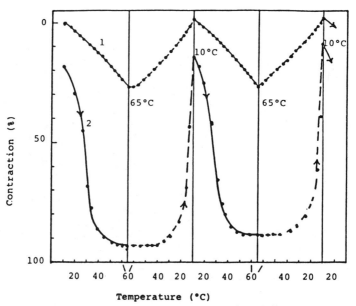

Figure 3   Temperature dependences of contraction of PMAA membranes with various
embedding fluids: (1) 70 mL of pure water, (2) 70 mL of 0.015 unit
mole/1 PEG solution.   Dry membrane: 10 mm-wide, 23-mm long, 4.7-mg
weight; load 490 mg; PEG molecular weight = 2000.   Ordinate is
expressed in % of the length of the dry membrane. [from Y. Osada,
J. Polymer Sci., Chem. Ed. 15, 255-267 (1977)]

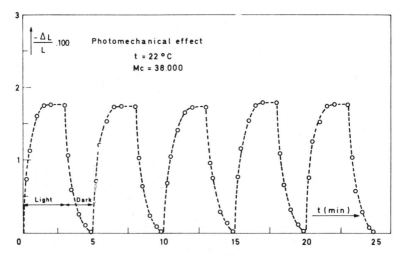

Figure 4   Contraction/dilation cycle of photochromic crosslinked poly(ethyl
acrylate).   Irradiation source, Hanau Q 700 Hg; l_0 = 3.97 cm; width
= 5 mm; load = 115 g; thickness = 0.48 mm; 1 = 84 mm; l_0 = ± 41 mm.
[from G. Smets, J. Polymer Sci., Polym. Chem. Ed., 13, 2223-2231 (1975)]

170

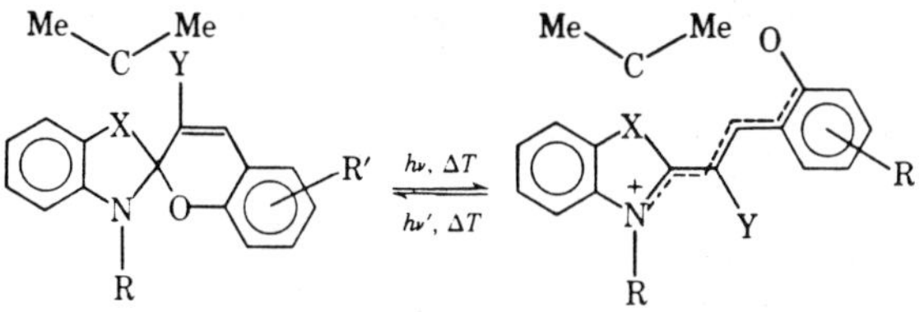

$h\nu$ = ultraviolet light, $h\nu'$ = visible light

When the indolinospirobenzopyran is converted into a divinyl compound and is used to crosslink poly(ethyl acrylate), a reversible expansion and contraction of the polymer takes place, as shown in Figure 4.

Polymers that Undergo Changes in Erosion Rate

An interesting erodible polymer system is based on polyacids that solubilize by an ionization of pendant carboxylic acid groups, a process highly dependent on the pH of the external environment. Well known examples are the partially esterified copolymers of methyl vinyl ether and maleic anhydride I (15-17).

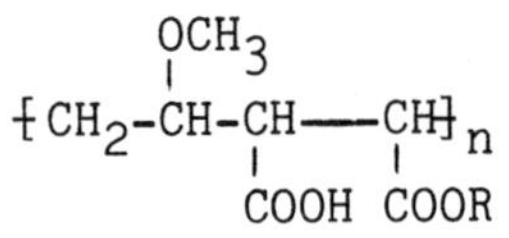

$$\left[CH_2-CH-CH\!\!-\!\!\!-CH \right]_n$$

with OCH_3 on the second CH, $COOH$ and $COOR$ on the last two carbons

I

These polymers were originally developed for enteric coating applications, which are designed to be stable up to a certain predetermined pH and then abruptly dissolve when that pH is exceeded. The pH at which the polymers dissolve depends on the size of the alkyl group in the ester portion of the polymer and, as shown in Figure 5 varies linearly with the number of carbons in the ester group (18).

These same polymers are also useful in controlled release applications. Thus, as shown in Figure 6, when a drug is homogeneously dispersed in polymer I and the polymer-drug composite is placed in a constant pH environment, a controlled dissolution of the polymer takes place with concomitant drug release (18).

171

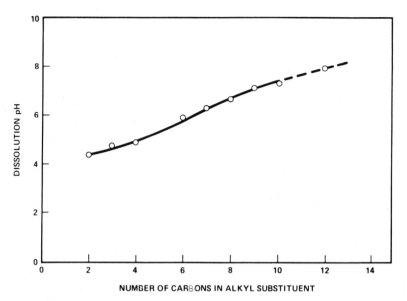

Figure 5   Relationship between pH of precipitation and size of ester group in
half-esters of methyl vinyl ether-maleic anhydride copolymers.
[from J. Heller, et al, J. Appl. Polym. Sci., 22, 1991-2009 (1978)]

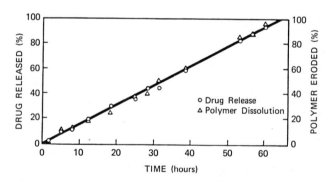

Figure 6   Rate of polymer dissolution and rate of release of hydrocortisone
for the n-butyl half-ester or methyl vinyl ether-maleic anhydride
copolymer containing 10 wt% drug dispersion. [from J. Heller,
et al, J. Appl. Polym. Sci., 22, 1991-2009 (1978)]

172

Because the size of the alkyl group in the ester portion of polymer I determines the pH of dissolution (Figure 5) and because the rate at which the polymer solubilizes depends on the magnitude of the difference between the pH of solubilization and the pH of the external environment, rate of polymer dissolution and hence rate of drug release is highly pH dependent. The extraordinary dependence of erosion rate on external pH is illustrated in Figure 7. Thus, this polymer system can be used for zero order drug release from constant pH environments or in applications where it is desired to significantly change the rate of drug delivery in response to small changes in the external pH.

One such application is a self-regulated drug delivery device in which hydrocortisone is released from an implant in response to the presence of specific amounts of external urea (19). Although such a device has no obvious therapeutic application, it was, nevertheless, of interest to investigate such a system as a model for the possible development of more therapeutically relevant devices.

The experimental device consists of an n-hexyl half-ester of polymer I containing dispersed micronized hydrocortisone in a core surrounded by a hydrogel containing immobilized urease. In a medium of constant pH and in the absence of external urea, the hydrocortisone release rate was that normally expected for that polymer at the given pH. However, in the presence of external urea, ammonium bicarbonate and ammonium hydroxide are generated within the hydrogel, and the consequent pH increase accelerates polymer erosion and hence drug release. This effect is shown in Figure 8.

The effect is reversible. Figure 9 shows changes in hydrocortisone release for devices that are alternately placed in solutions containing urea and solutions containing no urea. The decrease in delivery rate is not abrupt, but instead shows a first-order dependence. The first-order dependence is a consequence of the relatively water-insoluble hydrocortisone being trapped in the hydrogel layer, from which it is released by slow diffusion when the device is placed in solutions containing no urea.

Because this methodology is applicable to only a few cases in which an enzyme-substrate interaction produces a change in pH, another approach is currently under development (20). The essence of that approach is based on the homogeneous enzyme immunoassay developed at the Syva Company, as illustrated in Figure 10 (21). A selected trigger molecule (labeled as "drug" in Figure 10) is chemically bound to an enzyme close to the enzyme active site, and an antibody to the trigger molecule ("hapten") is then complexed

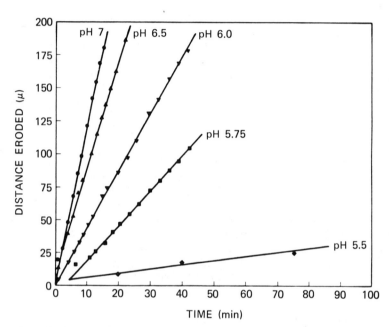

Figure 7   Effect of pH on erosion of half–esters of methyl vinyl ether–maleic anhydride copolymers. [from J. Heller, et al, J. Appl. Polym. Sci., _22_, 1991–2009 (1978)]

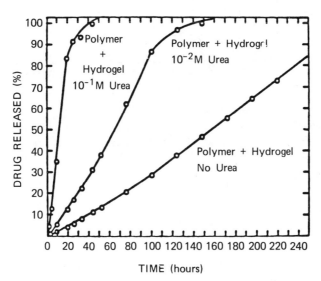

Figure 8 Hydrocortisone release rate at 35°C from a n-hexyl half-ester of a copolymer of methyl vinyl ether and maleic anhydride at pH 6.25 in the absence and presence of external urea. [from J. Heller and P.V. Trescony, J. Pharm. Sci., <u>68</u>, 919–921 (1979)]

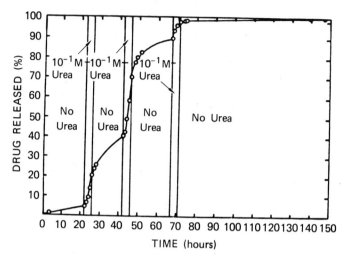

Figure 9   Hydrocortisone release rate at 35° from a n–hexyl half–ester of a copolymer of methyl vinyl ether and maleic anhydride at pH 6.25 as a function of sequential addition and removal of 10^{-1} M urea. [from J. Heller, et al, J. Appl. Polym. Sci., 22, 1991–2009 (1978)]

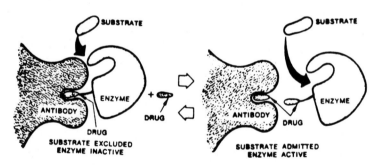

Figure 10   Principle of homogeneous enzyme immunoassay.   [from R.S. Schneider, et al, Clin. Chem., 19, 821–825 (1973)]

176

with the enzyme-trigger molecule conjugate. The antibody to the trigger molecule is obtained by sensitizing animals by injection of trigger molecules conjugated to a polymer and isolating antibodies from the blood serum. When the antibody complexes with the chemically bound trigger molecule, the active site of the enzyme is sterically shielded so that the substrate is excluded from the enzyme active site and the enzyme is rendered inactive. However, when free trigger molecules interact with the complex, they compete for the antibody binding sites and are able to displace some trigger molecule-enzyme conjugates from the complex, thus liberating active enzymes from an enzymatically inactive complex.

This methodology can be used in drug delivery devices by surrounding a bioerodible, drug-containing polymer with a hydrogel that contains immobilized inactivated enzymes. When activated, these enzymes are capable of degrading the hydrogel to small, water-soluble fragments, thus exposing the bioerodible polymer to the external environment.

The bioerodible polymer is constructed so that at pH 7.4, or at any other chosen pH, it bioerodes and releases its incorporated therapeutic agent at the desired rate; however, at a pH lower than that selected, no erosion or drug release takes place. The function of the hydrogel is to surround the bioerodible polymer with a low pH environment so that no polymer erosion takes place. However, when the hydrogel is removed by the activated enzyme, the polymer is exposed to the ambient pH and will begin to erode. Thus, the presence of any desired molecule in the external environment can initiate drug delivery by diffusing into the protective hydrogel and activating an enzyme.

Initial work with this delivery system used a lysozyme-morphine conjugate, which acted on a partially deacetylated chitin hydrogel (22). Current work uses an amylase-morphine conjugate acting on a starch hydrogel.

ACKNOWLEDGMENT

Work on self-regulated bioerodible drug delivery devices was partially supported by the World Health Organization under Grant No. H9/181/364 and by the National Institutes of Health under Grant No. GM27164.

REFERENCES

1. Kuhn, W.; Hargitay, B.; Katchalsky, A.; and Eisenberg, H. Nature, 1950, 165, 514-516.

2. Kuhn, W.; and Hargitay, B. Experientia, 1951, 7, 1-40.

3. Kuhn, W.; and Thurkauf, M. Kolloid-Z. v. Z. Polymere, 1962, 184, 114-117.

4. Kuhn, W.; Ramel, A.; Walters, D. H.; Ebner, G.; Kuhn H. J. Fortschr. Hochpolym.-Forsch, 1960, 1, 540-592.

5. Basu, S.; Chandhury, P. R. J. of Colloid Sci., 1957, 12, 19-24.

6. Bailey, F. E., Jr.; Lundberg, R. D.; Callard, R. W. J. Polymer Sci., 1964, A2, 845-851.

7. Osada, Y.; Saito, Y. Makromol. Chem. 1975, 176, 2761-2764.

8. Osada, Y.; Sato, M. J. Polymer Sci., Polym. Letters Ed., 1976, 14, 129-134.

9. Osada, Y. J. Polymer Sci., Polym. Chem. Ed., 1977, 15, 255-267.

10. Osada, Y. J. Polymer Sci., Polym. Chem. Ed., 1979, 17, 3485-3498.

11. Osada, Y. J. Polymer Sci., Polymer Lett. Ed., 1980, 18, 281-286.

12. Osada, Y.; Sato, M. Polymer, 1980, 21, 1057-1060.

13. Osada, Y.; Takeuchi, Y. J. Polymer Sci., Polymer Lett. Ed., 1981, 19, 303-308.

14. Smets, G. J. Polymer Sci., Polymer Chem. Ed., 1975, 13, 2223-2231.

15. Lappas, L. C.; McKeehan, W. J. Pharm. Sci. 1962, 51, 808.

16. Lappas, L. C.; McKeehan, W. J. Pharm. Sci., 1965, 54, 176-181.

17. Lappas, L. C.; McKeehan, W. J. Pharm. Sci., 1967 56, 1257-1261.

18. Heller, J.; Baker, R. W.; Gale, R. M.; Rodin, J. O. J. Appl. Polymer Sci., 1978, 22, 1991-2009.

19. Heller, J.; Trescony, P. V. J. Pharm. Sci., 1979, 68, 919-921.

20. Heller, J.; Pangburn, S. H.; Trescony, P. V., work in progress.

21. Schneider, R. S.; Lindquist, P.; Wong, E. T.; Rubenstein, K. E.; Ullman, E. F. Clin. Chem. 1973, 19, 821-825.

22. Pangburn, S. H.; Trescony, P. V.; Heller, J. Biomaterials, 1982, 3, 105-108.

MECHANISM OF THE BIODEGRADATION OF POLYCAPROLACTONE

P. Jarrett, C. V. Benedict, J. P. Bell,
J. A. Cameron and S. J. Huang

Institute of Materials Science
University of Connecticut
Storrs, Connecticut 06268

INTRODUCTION

The preparation of synthetic biodegradable polymers is an area of increasing interest (1-4). Disposal of nondegradable synthetic polymers after use is a problem that is growing more serious every day. The disposal of biodegradable polymers, on the other hand, is less difficult. Moreover, biodegradable polymers are becoming vital in the preparation of surgical implants, sutures, controlled release formulations of drugs and agricultural chemicals, weed suppressant covering, and mulches. Our efforts have been directed toward, a) the development of biodegradation testing methods, b) studying the factors affecting the biodegradation of synthetic polymers and the mechanisms of biodegradation, c) design and synthesis of new biodegradable polymers for special applications. During the course of our research to date we have found that both the nature of the chemical structures of the polymers and the morphology of the polymer samples affect the rates and extents of biodegradation of the polymer samples.

Among the biodegradable polymers aliphatic polyesters have received increasing attention as materials for sutures and as matrix materials for controlled released formulation of drugs and agricultural chemicals (5-7). We have studied in detail the microbial and enzymatic degradation of polycaprolactone (PCL), a polymer showing great promise as controlled release formulation matrix material. We report here our recent results.

EXPERIMENTAL

Microbial and enzymatic degradation methods for PCL films were recently reported (8-10). The degree of degradation by a yeast, _Cryptococcus laurentii_, and a fungus, _Fusarium_, was determined qualitatively in most cases. Surface effects of the organisms, on those samples which were still partially intact, were studied using a Cambridge Stereoscan (IIA) tungsten emission scanning electron microscope (SEM). The specimens were cut from the samples and fixed to the SEM stage using silver paint on the edges. All samples were dried and coated with approximately 100 Å gold before viewing in the SEM.

Crosslinked Samples

Samples of crosslinked polymer were prepared from PCL-700 (polycaprolactone-700), a product of Union Carbide reported to have $\bar{M}_w$ ~40,000. Solutions were prepared by dissolving the polymer pellets in benzene at room temperature. To these solutions were added varying amounts of recrystallized benzoyl peroxide (BPO). The weight percents of BPO in the polymer were 5%, 10%, 20%, 30% and 40%. Once the pellets and BPO were completely dissolved and the solution well mixed, it was poured into a flat bottomed, level petri dish upon a hot plate set to 75°C. The petri dish was then covered with a glass dome which sat upon a rubber gasket. Entrance and exit holes were cut in the gasket to allow the flow of nitrogen gas over the reacting mixture. This method was a result of efforts to prevent foaming due to the gases evolved during the BPO radical formation and reaction process. The solution was left at 75°C overnight to allow the reaction to go to completion. During this period the benzene evaporated completely, leaving a bubble-free, flat, crosslinked film. The film was then cut into sections which were put in a soxhlet extractor. They were extracted with benzene to remove all residues from the reaction as well as any non-crosslinked polymer. The films were then dried overnight in a vacuum oven.

The crosslinked films were prepared for the biodegradation study by cutting the films into 1 cm squares, .013 cm thick. The films were preweighed on an analytical balance. Five samples of each film were prepared in this way (two organisms were used, samples were run in duplicate, and there was a control sample). For a comparison the same tests were done on a melt crystallized film of non-crosslinked polymer.

Each sample was placed in a test tube containing a basal minimum salt solution (BMS) with 0.1% casamino acids as a growth stimulant. The BMS solution consisted of 0.7g K_2HPO_4, 0.7g $MgSO_4$ · $7H_2O$, distilled water at pH 6.4. Two test tubes for each degree of crosslinking were innoculated with the yeast, <u>Cryptococcus</u>, and the fungus, <u>Fusarium</u>. The samples were incubated at 25°C for 1 month with very mild agitation. After this period, intact films were removed for weighing and microscopy, and the remaining solution was freeze dried so that degradation products could be extracted and analyzed by gel permeation chromatography (GPC).

The approximate degree of crosslinking of the films was estimated by modulus determination above T_g and T_m using an Instron tensile test machine with oven at 68°C. The molecular weight between crosslinks (M_c) was estimated using rubber elasticity theory (11).

Polycaprolactone Single Crystal Experiments

Solution grown crystals of polycaprolactone were made by a mixed solvent precipitation process involving addition of an acetone solution containing the polymer to water at room temperature, to yield a fine suspension of polycaprolactone (PCL) particles in an aqueous medium. The degrading enzyme system was prepared by harvesting the supernatant of a culture of <u>Cryptococcus</u> <u>laurentii</u>, a yeast, which was grown in a Basal Minimum Salt solution containing 0.4% Casamino acids. The amount of degradation was followed by turbidity decrease.

RESULTS AND DISCUSSION

The susceptibility of a polymer to microbial attack depends on a variety of various factors. In addition to the nature of the chemical structure of the polymers, the morphology of the polymer samples greatly affects the rate of the biodegradation. Using scanning electron microscopy, we have found that the degradation of a partially crystalline polycaprolactone sample proceeds in a selective manner, with the amorphous regions being degraded prior to the degradation of the crystalline regions. Extracellular activites of the fungi are responsible for the observed biodegradation. (8-10) The fungi studied are producing an exoenzyme responsible for the biodegradation of the amorphous areas prior to the degradation of the crystalline spherulites. The sequence is attributable to the less ordered arrangement of the amorphous regions, allowing penetration of the enzyme into the polymer sample. The size, shape and number of the crystallites has a pronounced effect on the chain mobility in the amorphous regions, which has a controlling effect on biodegradation rate. The crystallites are ultimately degraded from the edges inward. These and other considerations will be discussed below.

Crosslinking of polymer films is generally accompanied by a reduction in crystallinity. PCL films were chemically crosslinked

183

with benzoyl peroxide and subjected to microbial degradation. The degradation was examined with SEM. It is clear in comparing Figures 2 and 3 to Figure 1 that the morphology of the partially degraded crosslinked films is quite different from that of the partially degraded non-crosslinked polymer. It appears that the crosslinked samples have many voids where degradation has taken place, leaving a sponge-like appearance. The non-crosslinked polymer shows a more uniform degradation, though a selectivity of amorphous over crystalline regions is evident. Supporting this data are two other observations. The degraded, highly crosslinked, films had the same dimensions as they did before degradation even though they lost ∿70% of their weight, while the non-crosslinked samples did not hold their dimensions. Also the 40% BPO sample showed no crystallinity by DSC data yet appeared slightly cloudy at room temperature. This latter observation is thought to be due to the presence of small voids formed in the polymer network matrix upon densification when cooled from the reaction temperature. The holes in the SEM micrographs could therefore be the result of the infiltration of these small voids by the degrading enzyme. This vast increase in internal surface area would explain why the outer dimensions of the films were not changed. Accompanying the crosslinking is a corresponding decrease in the degree of crystallinity. Originally, it was thought that this decrease in crystallinity would cause a major increase in the degradability of the polymer. The crosslinks themselves, however, appear to have a large effect on degradability of their own, as the results contained therein disclose, Table 1.

It is also clear that in degrading the network polymer (note the complete degradation of the 5% and 10% BPO samples), both organisms demonstrated the presence of an endo-enzyme, an enzyme

Table 1. Degradation of Characterized Crosslinked PCL Samples After One Month of Exposure to Cryptococcus and Fusarium

%BPO	M_c	% Crystallinity*	Degree of Degradation, wt. loss	
			Crypt.	Fung.
0%	----	65%	100%	100%
5%	7000	61%	100%	100%
10%	6300	50%	100%	100%
20%	3300	36%	87%	>50%
30%	----	2%	77%	>50%
40%	5800	0%	73%	>50%

* from differential scanning calorimetry, heating rate = 20°C/min.

Fig. 1.   SEM (1080X) of non-crosslinked partially crystalline
           polycaprolactone film.

Fig. 2.   SEM (750X) of partially degraded, non-crosslinked,
           polycaprolactone film.

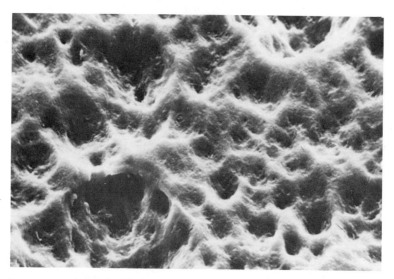

Fig. 3. SEM (500X) of partially degraded polycaprolactone 40%
BPO crosslinked sample degraded by <u>Cryptococcus</u>.

capable of cleaving a chain as opposed to an exo-enzyme which must
have a chain end available. The presence of exo-enzymes as well,
however, is not ruled out.

   Degradation of PCL crystals suspended in a solution containing
microorganisms was carried out. This was done in order to study
the attack of the degrading enzyme system at the PCL crystallite
level, removing the hindering effect of diffusion of enzyme into
the bulk material. The assumption was that the degradation mechanism would be similar for single crystals to that for crystallites
in the bulk material. In this experiment, the turbidity of the
suspension was measured every minute for ten minutes, showing the
degradation (loss of turbidity) with respect to time. The linearity of this plot indicates a pseudo first order kinetics process
is taking place, so that the slope of the line is proportional to
the activity of the enzyme. This relatively simple experiment
serves not only as an excellent assay for enzyme activity, but
also gives insight into the actual physical changes that a typical
polymer crystallite undergoes during biodegradation. The linearity of the plot is consistant with the hypothesis that attack on
the crystalline core material itself is limited to the edge of the
crystallite.

The edge attack hypothesis is one possible selective degradation mechanism. This hypothesis, as it relates to the turbidity experiment, should be discussed in order to better understand the turbidity data. It is thought that the degradation of the amorphous material on the top and bottom faces of the crystallite occurs most rapidly, until the crystalline core is reached. This change in thickness, however, should not affect the turbidity since the thickness dimension does not contribute significantly to the cross sectional area with respect to the incident beam. The initial change in thickness, therefore, from crystalline plus amorphous to crystalline material alone, need not be considered in the turbidity arguments. This aspect of the degradation process is being investigated with the use of small angle x-ray scattering (SAXS). We are assuming, then, that the particle approximates a disk that is changing only in radius. Any decrease in the mass of the crystallite will have a corresponding loss of area on the crystallite top and bottom faces. The linear decrease in turbidity therefore indicates that the degradation of the crystalline core is proceeding at a constant rate, and the slope of the turbidity line is proportional to that rate. The linearity of the turbidity versus time plot suggests that the accessible polymer substrate is in excess (since it appears to be zero order with respect to substrate). The SEM result, Fig. 4, of a partially degraded diamond shaped crystal supports this hypothesis.

Fig. 4.   SEM (1600X) of PCL crystal after exposure to Cryptococcus supernatant.

Turbidity data has also been used to show the presence of a non-enzymatic cofactor (Fig. 5). In this experiment, the enzymatic and non-enzymatic components of Cryptococcus laurentii supernatant were separated from each other. The two separated components showed little activity when run separately. Upon recombination of these two fractions, a synergistic effect is observed with respect to the relative activity. The exact nature of this cofactor is not known at the present time and is the subject of further study.

Closer examination of the biodegradation process has involved the use of Gel Permeation Chromatography (GPC). Fig. 6 shows the GPC results of the degradation of a PCL-700 (Union Carbide, Mw = 40,000, Mn = 18,500) film. Note that the high molecular weight peak is simply decreasing in size while there is no discernable

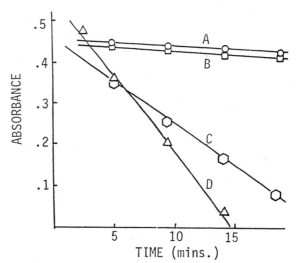

Fig. 5.   Turbidity change of crystal suspension during degradation of PCL by Cryptococcus.

A.   Heated supernatant of centrifuged culture – no enzyme present.
B.   Purified enzyme material – no cofactor present.
C.   Mixture of A and B, shows synergistic effect of enzyme with cofactor.
D.   Turbidity change of crystal suspension during degradation by Cryptococcus.

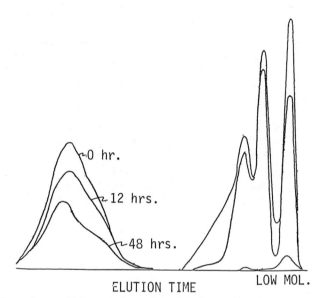

Fig. 6.   GPC results for biodegraded PCL film.

change in its position or distribution. In contrast to this, the data from the selective chemical degradation experiments (Fig. 7) show a shift toward lower molecular weight with degradation along with the eventual appearance of single and double traverse length peaks (length of chain needed to traverse the core thickness of a lamella). It was felt that the major difference here was not necessarily the degradation mechanism, but simply the inability of the enzyme system to diffuse into the polymer matrix, whereas the 40% methyl amine aqueous solution penetrates readily. It was for this reason that we decided to study the biodegradation of single crystals, to remove the diffusion factor. Comparison of Fig. 6 with Fig. 8, GPC results of the biodegradation of PCL-700 films and single crystals, shows a marked difference in the way the molecular weight is affected by degradation. The molecular weight shift visible in Fig. 8 is evidence of the presence of an endo-enzyme (an enzyme which cleaves the chain in a random fashion) in the degradation system. This was impossible to detect by any other means attempted when bulk polymer was used. The rapid decrease in the area under the high molecular weight peak during the single crystal biodegradation process, however, coupled with the lack of any appearance of single or double crystallite traverse length peaks, indicates that this (the endo-enzyme) is not the only process occurring and that perhaps an exo-enzyme (degradation from chain ends toward the center of the chain) is also involved. Further inspection of Fig. 8 shows the growth of a trimer peak with degradation. It is thought that this is the product of the exo-enzyme degradation mechanism.

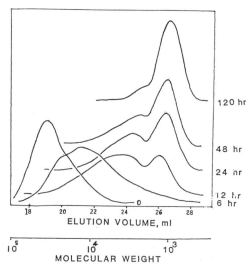

Fig. 7.   GPC results of PCL film degraded with 40% methylamine.

190

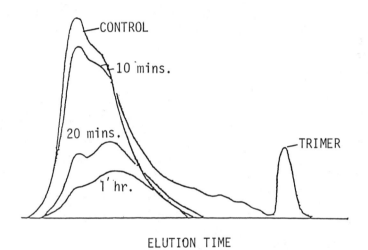

ELUTION TIME

Fig. 8.   GPC results for biodegraded PCL crystal suspension.

CONCLUSIONS

It has been shown that microorganisms tested produce both endo- and exo-enzymes and a cofactor (surfactant). These work together in the degradation of the polymers. Our present findings agree with our hypothesis that the chemical structure determines its biodegradability whereas the morphology and other physical properties affect the rate of biodegradation. In the absence of the cofactor the access of enzyme to the hydrophobic polymer is limited which explains the fact that very slow rates of degradation have been so far observed when polymer samples were incubated with purified enzymes.

ACKNOWLEDGEMENT

Financial support from the National Science Foundation (DMR80-13689) is gratefully acknowledged.

REFERENCES

1. S. J. Huang, M. Bitritto, K. W. Leong, J. Pavlisko, R. Roby and J. R. Knox, Advances in Chemistry Series No. 169. "Stabilization and Degradation of Polymers", D. L. Atlasa and W. L. Hawkins, Eds., Am. Chem. Soc., 1978, pp. 205-214.
2. J. E. Potts, in "Aspects of Degradation and Stabilization of Polymers", H. H. G. Jellinek, Ed., Elsevier, New York, 1978, pp. 617-657.
3. J. E. Guillet, in "Polymers and Ecological Problems", J. E. Guillet, Ed., Plenum Press, New York, 1973, pp. 1-26.
4. F. Rodriguez, Chem. Technol., 409 (1971).
5. A. Schinler and G. G. Pitt, Polym. Reprints 23, 111 (1982).
6. N. B. Graham, Brit. Polym. J., 10:260 (1978).
7. J. Heller, Biomaterials, 1:51 (1980).
8. W. Cook, J. A. Cameron, J. P. Bell and S. J. Huang, J. Polym. Sci.: Polym. Lett. Ed., 19:159 (1981).
9. C. V. Benedict, W. J. Cook, P. Janett, J. A. Cameron, S. J. Huang, and J. P. Bell, J. Appl. Polym. Sci., 28, 327(1983).
10. C. V. Benedict, J. A. Cameron, and S. J. Huang, J. Appl. Polym. Sci., 28, 335(1983).
11. J. J. Aklonis, et al., "Introduction to Polymer Viscoelasticity", Wiley Interscience, New York, 1972, p. 121.

SWELLING BEHAVIOR OF GLUCOSE SENSITIVE MEMBRANES

Thomas A. Horbett, Joseph Kost and Buddy D. Ratner

University of Washington
Department of Chemical Engineering
Seattle, Washington 98195

INTRODUCTION

Polymers which are sensitive to their environment may be useful as metabolic sensors. Examples of environmentally sensitive polymers include partially hydrolyzed polyacrylamide gels that undergo phase transitions in response to small changes in pH, ionic strength or voltage (1,2), ethylene-vinyl-N,N,-diethylglycinate copolymers whose permeability is pH dependent (3), and methyl vinyl ether-maleic anhydride copolymers with entrapped urease that undergo accelerated erosion in response to urea (4). Because of the great need for a glucose sensor which could be used _in vivo_ to enhance treatment of diabetics, we are developing membranes responsive to glucose by incorporating glucose oxidase into amine containing polymers. The conversion of glucose into gluconic acid in the membrane causes increased protonation of the membrane resulting in increased membrane swelling. Figure 1 is a schematic illustration of how the glucose sensitive membrane is thought to operate. This paper describes the preparation and swelling properties of glucose sensitive membranes of this type.

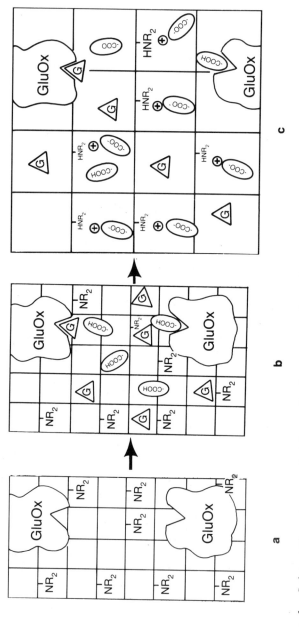

Figure 1. Schematic representation of the proposed mechanism of action of a glucose sensitive membrane (see "key" for these diagrams on the next page) (a) In the absence of glucose, at physiologic pH, few of the amine groups are protonated. (b) In the presence of glucose, the glucose oxidase produces gluconic acid which can, (c) protonate the amine groups. The fixed positive changes on the polymeric network lead to electrostatic repulsion and membrane swelling.

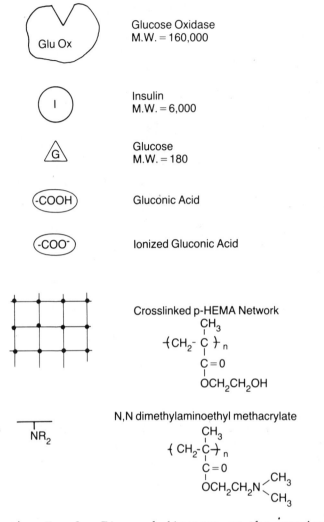

Glucose Oxidase
M.W. = 160,000

Insulin
M.W. = 6,000

Glucose
M.W. = 180

Gluconic Acid

Ionized Gluconic Acid

Crosslinked p-HEMA Network

$$\{CH_2\text{-}\overset{\overset{\displaystyle CH_3}{|}}{C}\}_n$$

$$C=O$$

$$OCH_2CH_2OH$$

N,N dimethylaminoethyl methacrylate

$$\{CH_2\text{-}\overset{\overset{\displaystyle CH_3}{|}}{C}\}_n$$

$$C=O$$

$$OCH_2CH_2N\overset{\diagup CH_3}{\diagdown CH_3}$$

Figure 1. (cont) - Key for Figure 1 diagrams on the previous page.

EXPERIMENTAL

The monomers used were 2-hydroxyethyl methacrylate (HEMA), N,N,-dimethylaminoethyl methacrylate (NNDMAEM), and tetraethylene glycol dimethacrylate (TEGDMA). Glucose oxidase (GO) type VII from Aspergillus niger (125,000 units per gram solid) was obtained from Sigma Chemical Co., St. Louis, Mo. The buffer used was 0.01 M citrate, 0.01 M phosphate, 0.12 NaCl, 0.02% sodium azide, pH 7.4 (CPBSz). Glucose was used in the form of dextrose.

Membranes were prepared with a low temperature, radiation initiated polymerization technique based on the work of Kaetsu et al. (5). HEMA, NNDMAEM, TEGDMA and ethylene glycol (EG) were mixed and then added to H_2O containing the GO. The quantities of each component for each membrane are shown in Table 1. The final mixture was poured between two glass plates separated by stainless steel shims. The assembly was frozen at -70°C in a freezer. It was then removed, sealed in a polyester bag, placed in a Dewar flask containing dry ice and acetone, and irradiated with 0.25 M Rad in a ^{60}Co source. After irradiation, the plates were separated by soaking in CPBSz in a refrigerator for several days. The membranes were stored in CPBSz at room temperature.

Water contents of membranes were determined gravimetrically, as previously described (6). They are calculated as the difference in weight between the wet and dry sample divided by the wet weight times 100. Each value plotted is the average water content measured for two specimens. The replicates agreed within 2%. pH was measured with a Radiometer PHM62 pH meter.

RESULTS AND DISCUSSION

The membranes contained a base polymer (HEMA), an amine component (NNDMAEM), and a cross-linker (TEGDMA). In terms of the monomer composition, the amine portion corresponds to 8.8, 17.5 or 35.5% by volume in the original monomer mixture (membranes E, F and G in Table 1). The GO concentration in the final mixture was 9.8 mg/ml in membranes E, F and G, but was 0.98 mg/ml in H and 19.6 in I.

The kinetics of swelling and water content increases were studied by using discs of equal size cut from a larger membrane with a cork borer. The GO membrane discs were placed in petri dishes containing water plus various amounts of glucose and kept in an oxygen atmosphere. A series of measurements were made at various times after putting the discs in the solutions. The measurements included pH of the buffer in the petri dish (Figures 2 and 3), the diameter of the disc (Figures 4 and 5), and the water content of the disc (Figures 6 and 7). Each of these pairs of figures

Table 1. Membrane Formulations[1]

Membrane	HEMA	NNDMAEM	TEGDMA	EG	H_2O	GO
E	5.0	0.5	0.2	0.5	4.5	105
F	4.5	1.0	0.2	0.5	4.5	105
G	3.5	2.0	0.2	0.5	4.5	105
H	5.0	0.5	0.2	0.5	4.5	10.5
I	5.0	0.5	0.2	0.5	4.5	210

[1] Figures given are volume in ml except GO, which is given in mg.

originated from the same data, but one shows the data plotted versus time and the other versus glucose concentration, a procedure found quite useful in analysis of the results, as will be seen.

The pH of the glucose solutions in which the immobilized GO discs were kept is plotted versus time of equilibration in Figure 2. No change in pH occured in the absence of glucose (top line), but pH decreased rapidly when glucose was present. The pH changes induced by very low glucose (10 mg %) were less than the pH decrease casued by physiologic glucose (100 mg %), while physiologically hyperglycemic glucose levels (500 mg %) induced still larger pH changes. The time required for achievement of the final pH also varied with glucose concentration.

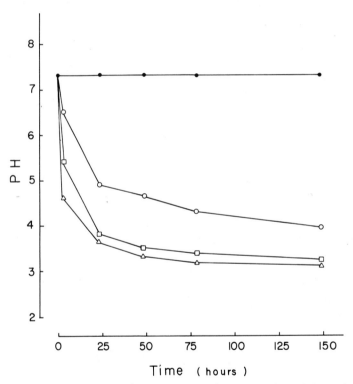

Figure 2.   Kinetics of bulk phase pH change induced by glucose sensitive membranes. Data were taken with membrane E exposed to the following initial glucose concentration (in mg %): 0 (solid circles), 10 (open circles), 100 (squares) or 500 (triangles).

Figure 3 shows the pH of the solutions plotted versus initial glucose concentration. The short term response (3 hours) varied markedly with glucose concentration. As gluconic acid was produced, within the membrane, the absolute concentration of acid in the outside solution increased, rendering the effect of further acid generation within the membrane less effective in changing the bulk phase concentration. That is, greater concentrations of acid have more "resistance" to pH changes and act like a buffer. Thus, at shorter times, the pH had not decreased as much as possible and further pH changes could occur more readily in response to further acid changes. The net effect is to render the system most responsive to glucose at shorter times, as observed.

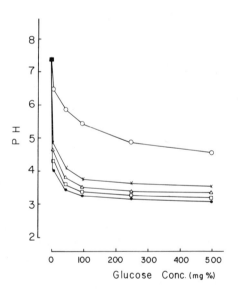

Figure 3. Effect of glucose concentration on bulk phase pH change induced by glucose sensitive membranes. The pH of solutions containing membrane E and glucose were measured at the following times (in hours): 3 (open circles), 24 (X's), 49 (triangles), 76.5 (squares), or 144 (solid circles).

The diameter of the membranes exposed to glucose solutions increased by approximately 20%, depending on the glucose concentration (Fig. 4). After 1 or 3 hours the diameter increased in 10 mg % glucose but did not further increase at higher glucose concentrations (Fig. 5). At 48 hours, all discs were somewhat larger than at shorter times, but, again, little difference in diameter of membranes exposed to various glucose concentrations was evident except at the lowest concentration (10 mg %).

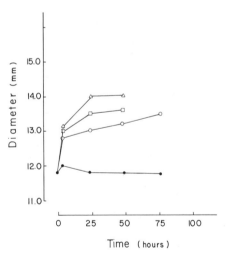

Figure 4. Kinetics of swelling of glucose sensitive membranes. The diameter of membrane discs of formulation E after various periods of exposure to glucose solutions are plotted. The glucose concentrations (in mg %) were: 0 (solid circles), 10 (open circles), 100 (squares), or 500 (triangles).

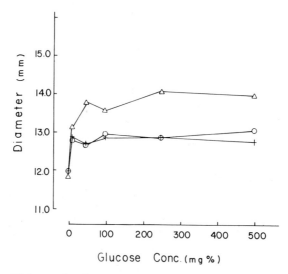

Figure 5. Swelling of glucose sensitive membranes in glucose solutions. The diameter of discs of membrane formulation E were measured after various periods of exposure to the stated glucose concentration. Data taken at the following times (in hours) are plotted: 1 (+), 3 (circles), and 48 (triangles).

The increase in water content of the membranes exposed to glucose solutions was qualitatively similar to the increases in diameter (Figures 6 and 7). Water content remained at about 60% in the absence of glucose, but increased to almost 75% at higher glucose concentrations and longer times. At the shortest time measured, 3 hours, a 16-18% increase in water content occurred. The water content of membranes after 45 min. or 3.5 hrs. (lower curves in Fig. 7) also showed little difference once at least 10 mg % glucose was present. At longer times, further water uptake occured, but again little difference was evident as a function of glucose concentration.

200

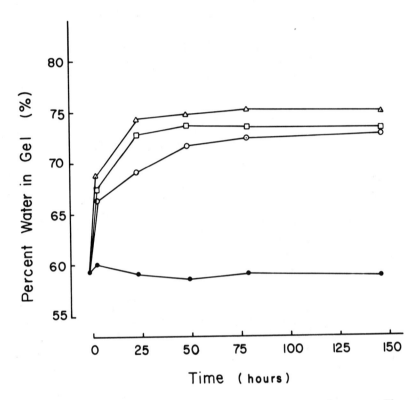

Figure 6. Water content of glucose sensitive membranes. The water content of membrane formulation E was measured after exposure to the following glucose concentrations (in mg %): 0 (solid circles), 10 (open circles), 100 (squares), and 500 (triangles).

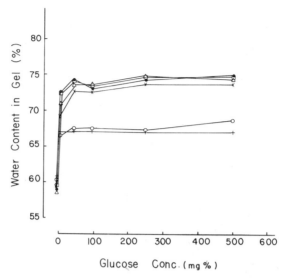

Figure 7. Water content of glucose sensitive membranes. The water content of membrane formulation E was measured after exposure to glucose for the following times (in hours): 0.75 (cross), 3.5 (open circles), 24.5 (X), 48.5 (triangles), 78 (squares) or 144 (solid circles).

The swelling and water content results revealed several interesting features of the immobilized GO membrane response to glucose. First, the membranes did, in fact, swell and become more hydrated in the presence of glucose. The membranes could therefore be used as glucose sensors, by coupling the swelling change to a force or pressure sensor. Visual inspection of the membrane easily detected the swelling: membranes in glucose were noticeably more swollen and more translucent than controls. The ready visual detection was undoubtedly due to the fact that the area of the membrane increases in proportion to the square of the radius, so that the approximately 20% change in diameter measured corresponded to a 44% change in area. A glucose detector system based on detecting area changes (or, in the ultimate case, volume) would be

202

much more sensitive to glucose than the diameter measurements so far made.

Analysis of the effect of glucose concentration on swelling or water content provides further insight into the properties of these membranes. Inspection of the pH changes induced by the membrane (Figure 2) reveals that at short times, the bulk phase pH was still much above the final values, especially at 10 mg % glucose where the 3 hour pH was about 6.5. Nonetheless, the membranes in 10 mg % glucose for 3 hours were swollen as much as membranes in all higher glucose concentrations (Figures 5 and 7). At longer times, swelling increased, but still varied little with glucose concentrations. It appears that two stages of swelling and water content increases can occur. In the first stage, the membrane swelling and water content are the result of effects due to the internal pH of the membrane and the bulk phase pH. The internal pH is probably lower than the bulk pH in this stage. The higher bulk phase pH probably restrains the membrane from swelling completely (lower curves in Figures 5 and 7). In the second stage, once bulk phase pH is also lowered, the membrane assumes the maximum swelling and water content values characteristic of a system in which internal and external pH are both quite acid (upper curves Figures 5 and 7). This analysis thus suggests that the membranes are able to assume internal pH values different that the bulk since otherwise the membrane in 10 mg % glucose (at 3 hours), where bulk pH is 6.5, should not be as swollen as a membrane in 500 mg %, where bulk pH at 3 hours is much lower, namely 4.5. This conclusion is important to the eventual use of these membranes in the presence of physiologically buffered fluids such as blood plasma since it indicates changes in the membrane (e.g., swelling or permeability increases) could still occur under these circumstances.

The fact that membranes in 10 mg % glucose are swollen to approximately the same degree as membranes in 500 mg %, glucose at 3 hours, led us to examine membranes with higher amine content since we desired the initial membrane response to vary with glucose concentration. The properties of membranes E, F, and G containing 1x, 2x, or 4x amine content, respectively, are shown in Figure 8 and 9. In Figure 8, the increase in water content induced by the changes in pH which occur during glucose turnover are cross-plotted. The initial water content (at pH 7.4) varied with amine content. The increase in water content (16 - 18%) caused by the pH decrease was about the same on all membranes. The swelling of membranes without any GO in response to buffers of differing pH was very similar to the glucose induced swelling data (7).

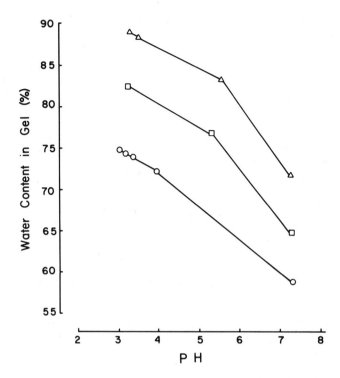

Figure 8. Effect of pH on the swelling of HEMA–NNDMAEM–GO Membranes. Membranes E (circles), F (squares), and G (triangles) were evaluated.

Figure 9 shows the effect of glucose concentration on water content of these same membranes. Normal blood glucose levels are 100 mg %, while diabetics can be as high as 500 mg %. In this Figure the steady state swelling reached after relatively long times (ca 24 hours) is shown. Little difference in swelling at various glucose concentrations was evident in the membrane with lowest amine content (E). Membranes containing higher amine contents (F and G) did show a differential response to glucose concentrations in the physiologic range. These results suggest that appropriately formulated membranes have the potential to discriminate physiologically relevant glucose concentrations and may possibly be useful in glucose detection.

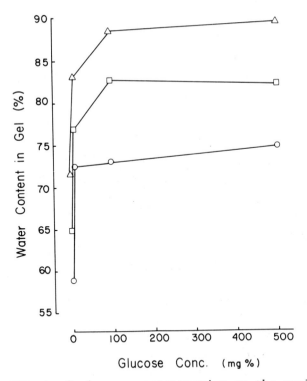

Figure 9. Effect of glucose concentration on the swelling of
HEMA-NNDMAEM-GO membranes. Membranes E (circles), F (squares),
and G (triangles) were evaluated.

To determine the sensitivity of such membranes under conditions more closely simulating the in vivo environment, swelling
measurements were also made on membranes kept in 0.15 M NaCl
solutions. As shown in Figure 10, the degree of swelling of
membrane E in response to glucose was only about 3%, in contrast to
the 16% change seen in pure water (see Figure 9). However, by
adding additional GO to the formulation (membrane I), the swelling
increase in response to glucose was enhanced to about 6%. The
decreased swelling in the presence of salt is undoubtedly due to
the reduction in electrostatic repulsion by counter ion shielding.
The enhanced swelling in saline induced by increased GO may be due
to a change in the mechanical properties of the membrane or an
increased rate of glucose turnover in the domain of the membrane.
We have more recently found that glucose induced swelling of
membranes F and G (containing higher amine concentrations) in

saline was much greater than observed for membrane E (7). The results show the flexibility inherent in this polymer system in achieving suitable properties for glucose detection.

The HEMA-NNDMAEM-GO membranes have thus been shown to be sensitive to glucose concentrations in the physiologic range. Their ability to respond to glucose variations under truly physiologic conditions (e.g., with constant external pH) remains to be demonstrated. The flexibility inherent in the monomer formulations and the demonstrated effect of such changes on membrane properties suggest that a successful in vivo glucose sensor based on these membranes may be an achievable goal. The use of these membranes to

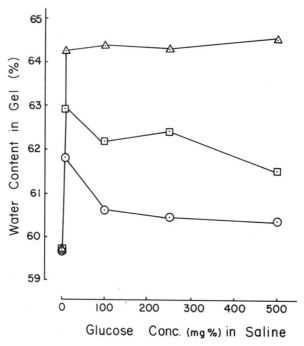

Figure 10. Effect of glucose concentration on the swelling of HEMA-NNDMAEM-GO membranes in saline. Membranes H (circles), E (squares), and I (triangles) were evaluated. The water content was measured after several days of equilibrium in glucose.

control insulin delivery in response to glucose concentrations is also under active investigation (7,8).

REFERENCES

1. T. Tanaka, Gels, <u>Sci. Am.</u> 244:124 (1981).

2. T. Tanaka, I. Nishio, S.T. Sun and S. Ueno-Nishio, Collapse of gels in an electric field, <u>Science</u> 218:467 (1982).

3. F. Alahaique, M. Marchetti, F.M. Reccieri and E. Santucci, A polymeric film responding in diffusion properties to environmental pH stimuli: a model for a self-regulating drug delivery system, <u>J. Pharmacol.</u> 33:413 (1981).

4. J. Heller and P.V. Trescony, Controlled drug release by polymer dissolution II: enzyme-mediated delivery device, <u>J. Pharm. Sci.</u> 68:919 (1979).

5. I. Kaetsu, M. Kumakura and M. Yoshida, Enzyme immobilization by radiation-induced polymerization of 2-hydroxyethyl methacrylate at low temperatures, <u>Biotechnol. Bioeng.</u> 21: 847 (1979).

6. B.D. Ratner and I.F. Miller, Interaction of urea with poly(2-hydroxyethyl methacrylate) hydrogels, <u>J. Polym. Sci.</u> Part A-1 10:2425 (1972).

7. J. Kost, T.A. Horbett, B.D. Ratner and M. Singh, Manuscript submitted, 1983.

8. T.A. Horbett, B.D. Ratner, J. Kost and M. Singh, A bioresponsive membrane for insulin delivery. In: Proceedings of the International Symposium on Recent Advances in Drug Delivery Systems, Plenum Press, Plenum N.Y., in press. 1984.

ACKNOWLEDGEMENTS

The financial assistance of the University of Washington Graduate School Research Fund, the California and Washington Affiliates of the American Diabetes Association, and the National Institute of Arthritis, Diabetes, and Digestive and Kidney Diseases (Grant No. AM30770) is gratefully acknowledged.

SELECTED ASPECTS OF CELL AND MOLECULAR BIOLOGY OF IN VIVO BIOCOMPATIBILITY

Roger E. Marchant, Kathleen M. Miller,
Anne Hiltner and James M. Anderson

Departments of Pathology and Macromolecular Science
Case Western Reserve University
Cleveland, Ohio 44106

INTRODUCTION

The biocompatibility of an implanted material or prosthetic device is a dynamic and two-way process that involves the time dependent effects of the host on the material and the material on the host. The implantation of any synthetic material initiates a wound healing mechanism that is characterized by the inflammatory response. However, the inflammatory response itself involves complex and highly regulated interactions between specific cells and various molecular mediators. An understanding of these interactions which occur following biomaterial implantation has been hindered by the difficulty in quantifying the cellular and biological events.

As part of our program to develop a better understanding of the phenomenon which leads to the biocompatibility or the biodegradation of materials, an in vivo model system has been developed to quantitatively and qualitatively characterize biological changes in the acute and chronic inflammatory response that occur due to polymer implantation. This system permits the serial evaluation of the cellular and humoral components of the exudate which initially surrounds the polymeric material following implantation and ultimately, provides access to monitor the degradation of implant materials relative to the concomitant biological changes. Materials which have been studied using the cage implant system include a biodegradable hydrogel poly(2-hydroxyethyl-L-glutamine)[1], Biomer®[2], Nylon 6, expanded polytetrafluoroethylene and silicone rubber[3]. Figure 1 outlines the various types of studies that may be performed utilizing the cage implant system. Some of the results from these studies will form the basis of a discussion directed toward appreciating the role that inflammation, complement, chemotaxis and frustrated phagocytosis may play in determining the biocompatibility of a material.

209

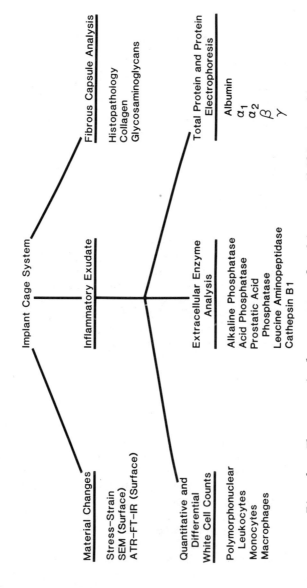

Fig. 1. The cage implant system for biocompatibility studies.

The Acute Inflammatory Response

Inflammation is the host's response to injury or the presence of injurious agents. The surgery of implantation induces the initial reactions which include a series of interdependent events that begin with hemodynamic changes, followed by alterations in vascular permeability. The increased permeability of the adjacent microvasculature promotes the transport of protein-rich inflammatory fluid or exudate into the extravascular tissues and wound site. Simultaneously, circulating leukocytes interact with stimulatory factors, adhere to the blood vessel endothelium and subsequently pass through the vessel wall and into the extravascular tissue around the implant. Figure 2 illustrates the wound healing response during an inflammatory reaction to an implanted material. The intensity and duration of the response is controlled by a variety of mediators and determined by the size and nature of the implanted material, the site of implantation and the reactive capability of the host. A brief cellular response of low intensity would be indicative of tissue compatibility.

The acute phase of inflammation is characterized by the preferential migration of one particular leukocyte, the neutrophil (a polymorphonuclear leukocyte (PMN)). The peak migration usually occurs within the first 72 hours following injury[4]. Mononuclear leukocytes (macrophages and lymphocytes) predominate at later stages of the inflammatory process. Macrophages are derived from blood monocytes. Once monocytes leave the circulation and enter the tissue they undergo a transformation in response to various stimuli and differentiate into macrophages. It is now accepted that PMNs, monocytes and, to a lesser extent, lymphocytes are drawn from the vasculature to the site of injury by locally generated chemotactic factors.

The transport of plasma components, inflammatory mediators and leukocytes into the area of injury leads to a fluid phase or exudate which can be monitored using the cage implant system. In the cage implant system the biomaterial of interest is placed inside a stainless steel cylindrical cage (3.5 cm in length and 1 cm in diameter)[1], which is then implanted subcutaneously in the back of a rat. Small aliquots of the inflammatory exudate which surrounds the biomaterial can be periodically aspirated using a syringe and analyzed. This system then offers the opportunity to quantify the variations in different leukocyte concentrations (PMNs, macrophages and lymphocytes), cell function, and other exudate constituents during the inflammatory and early healing response to the implanted material within the cage. The results may then be statistically compared to the results for an implanted cage without polymer or one containing a negative control. An example of these data is provided by Table 1, which shows the variation in PMN concentration in the exudates of three materials with implantation time. The table shows that the implanted polymers did not significantly elevate the exudate PMN concentration, when statistically (95% confidence level) compared to the control implant with-

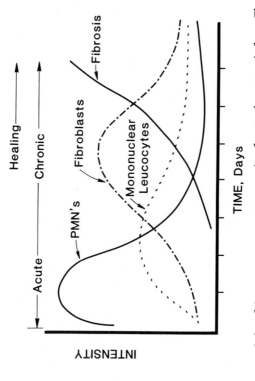

Fig. 2. The wound healing response to an implanted material. The intensity and duration of the response will be determined by the size and nature of the implanted material and by host factors.

Table 1. The variation in polymorphonuclear leukocyte concentration in the exudate as a function of implantation time[3].

Time	PMNs per µl			
(days)	control	PHEG	Biomer®	Silicone
4	3500±310	3070±370	4830±820	3480±370
7	390±110	280± 40	770±230	480±210
14	12± 4	19± 6	20± 9	32± 34

Mean Value±Standard Deviation
Biomer® samples were obtained from Dr. J. Kardos of Washington University, St. Louis, MO.
PHEG: poly(2-hydroxyethyl-L-glutamine).

out a polymer. This suggests that cytotoxic extractables were not being released into the exudate by these materials. However, both PHEG and Biomer® appear to have a significant effect on other aspects of the inflammatory response particularly during the chronic phase of inflammation. A comprehensive analysis of the exudate derived from these implants has been reported[1-3].

Phagocytosis and Cellular Activation

Phagocytosis refers to the ability of phagocytic cells (i.e., PMNs and macrophages) to engulf and digest particles. The particles are normally represented by cellular or tissue debris, bacteria or other extraneous matter present in the inflammatory region. In addition to being longer-lived than the PMN, the macrophage is capable of synthesizing many proteins that affect the inflammatory response such as complement factors. Once the phagocyte recognizes the particle, attachment followed by engulfment occurs. The cell extends sections of its cytoplasm (pseudopods) around the object to be engulfed and eventually completely encloses the particle within the cytoplasmic membrane. The limiting membrane of this phagosome fuses with the limiting membranes of enzyme-rich lysosomal granules, resulting in the discharge of the granule contents into this vesicle now called a phagolysosome. The phagolysosome is uniquely suited, with its low pH and high concentration of destructive enzymes of broad specificity, to effect the complete degradation of most ingested particles.

At the initiation of phagocytosis the cell becomes metabolically activated and a respiratory burst follows which leads to the production of hydrogen peroxide, oxygen radicals and anions. These short-lived and highly reactive species are effective killers of many types of bacteria and provide a potential source of initiators for the degradation of addition polymers. During the process of phagocytosis there may be some leakage (exocytosis) of lysosomal enzymes and metabolic products (e.g., H_2O_2) from the phagocyte into the external medium or exudate.

In the presence of a biomaterial implant, activation of the inflammatory cells is considered to occur following the adhesion of the
cells to the polymer surface or through a nonadhesive mechanism where
phagocytes in the exudate are activated through cell-cell interactions
or cell-mediator interactions[1]. Figure 3 shows a scheme by which
cells in the exudate can interact with a protein-coated biomaterial
surface. For macrophages, cellular activation by adhesive and nonadhesive events can lead to the presence of extracellular lysosomal
enzymes through non-cytolytic release by an exocytosis mechanism or as
a result of necrosis with lytic release. For PMNs, however, the importance of cell-surface adhesion with respect to the exocytosis mechanism has been shown by Gallin[5] and Henson[6]. The studies of Wright
and Gallin[5] have shown that the extracellular release of the contents
of specific granules in PMNs were significantly augmented when the
cells actively adhered to polymeric surfaces.

A biomaterial implant represents a particle which PMNs and macrophages are unable to completely engulf. This can lead to the incom-
plete fusion of the phagolysosomes with the plasma membrane and thus
to the extracellular release of enzymes by exocytosis. This process
has been referred to by Henson[6] as frustrated phagocytosis. Henson
suggested that the specific mode of cell activation in the inflammatory response is dependent on the size of the implant and that a material in powder or particulate form which is suitable for phagocytosis may provoke a different degree of inflammation than the same
material in a nonphagocytosable form such as a film, where frustrated
phagocytosis and exocytosis will probably prevail.

Frustrated phagocytosis and the mechanism of exocytosis can be
used to explain the elevated levels of alkaline phosphatase (derived
from specific granules of PMNs) and acid phosphatase (derived from
PMNs and macrophages) which we have observed in the exudates around
PHEG and Biomer® implants when compared to silicone implants and
empty cage controls[3]. Lysosomal enzyme release will be strongly
affected by the consequences of these cell-surface interactions. In
our studies, light microscopic examination of the retrieved polymer
samples have demonstrated that inflammatory cells do actively adhere
to PHEG and Biomer® surfaces. Figure 4 illustrates this phenomenon
where PMNs and macrophages can be seen adherent to the surface of a
Biomer® implant at 4 days postimplantation time.

Mediators of Inflammation

Central to the molecular biology of inflammation is the complement system. The complement system which is a complex and extensive
series of proteins and glycoproteins can be activated by antigen-
antibody complexes, bacterial polysaccharides, endotoxins, proteolytic enzymes and numerous synthetic polymers such as Nylon 6, cellophane, polymethylmethacrylate and Dacron®. Once initiated, activation proceeds along either the classical or alternative pathways

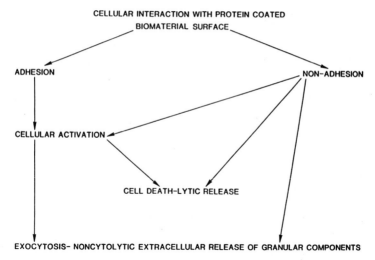

CELLULAR INTERACTION WITH PROTEIN COATED
BIOMATERIAL SURFACE

ADHESION

NON-ADHESION

CELLULAR ACTIVATION

CELL DEATH-LYTIC RELEASE

EXOCYTOSIS- NONCYTOLYTIC EXTRACELLULAR RELEASE OF GRANULAR COMPONENTS

Fig. 3. The interaction of phagocytes with a biomaterial.

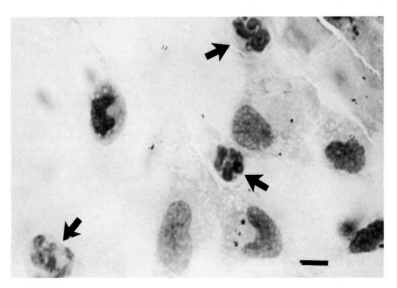

Fig. 4. Optical micrograph. PMNs (arrows) and macrophages adherent to the surface of a Biomer® implant at 4 days postimplantation time. The dimension bar represents 10μm.

215

through a series of cascading enzymatic reactions some of which are indicated in Figure 5. The reactions of the complement cascade generate numerous biological effects which influence much of the inflammatory process including vascular permeability, leukocyte chemotaxis, stimulation of oxidative metabolism, opsonization of particles, leukocyte adhesion, attachment and enzyme release, and membrane cell lysis.

The C3 component which is of pivotal importance to the complement system has the highest plasma concentration of any factor and is known to be present in the inflammatory exudates around polymers in the cage implant system[7]. Enzymatic cleavage of C3 results in the generation of the C3a and C3b fragments. Adsorption of C3b to surfaces renders a particle of opsonized or recognizable to phagocytes and thus promotes cellular adhesion. In the presence of factor D and magnesium ions surface bound C3b is activated by factor B of the alternative pathway to create the enzyme complex C3bBb which, when stabilized, cleaves C3 into C3a and C3b providing a feedback mechanism for the system. The C3bBb complex can spontaneously dissociate before being stabilized to release fragment Bb which has been shown to enhance the adherence and spreading of macrophages over a surface[8]. C3b also cleaves C5 into C5a and C5b.

The small fluid phase fragments C3a and C5a have important roles in the inflammatory process. They regulate the level and intensity of the inflammatory response by inducing smooth muscle contraction, enhancing vascular permeability, inducing the release of vasoactive amines from basophils and mast cells and inducing lysosomal enzyme release from PMNs. In addition, C5a is a potent chemoattractant and as such serves to direct the responsive leukocytes, PMNs and macrophages. Membrane perturbation of macrophages and PMNs by receptor-bound C5a also induces lysosomal enzyme release and degranulation[9,10]. Lysosomal enzymes are not only capable of causing increased local tissue damage at the sites of inflammation but also cleave C5 to yield C5a which in turn can further perturb the level of inflammation.

The macrophage is probably the most highly regulatory cell associated with the inflammatory response, because of its de novo synthesis abilities and the fact that it remains available and active for a long period. As Figure 6 shows, the multiple macrophage functions have broad effects. In addition to complement associated regulation of the inflammatory response, the macrophage through the secretion of mediators, regulates inflammatory cells like PMNs and fibroblasts, which, in turn, will generate soluble products that affect the macrophage (see Figure 6). Consequently, the fate of the macrophages following surface interactions with a biomaterial will strongly influence the level of inflammation and thus, the material's degree of tissue compatibility.

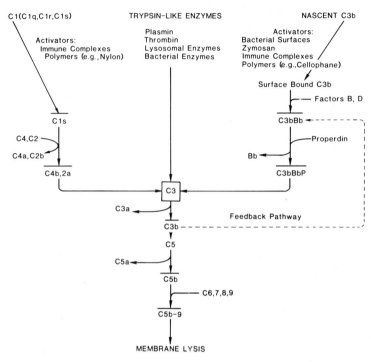

CLASSICAL PATHWAY   OTHER   ALTERNATIVE PATHWAY

Fig. 5.   Pathways of complement activation with generation of frag-
          ments central to the regulation of the inflammatory response.

Healing and Repair

      The localized destruction of tissue in an inflammatory site
places a physical separation between areas of healthy tissue. If possible, the injured area will be replaced by the regeneration of lost
tissue in order to restore its previous anatomical and functional
condition. More generally, however, and almost certainly in an area
around a biomaterial implant, the injured area will heal by connective
tissue repair (scarring).

      Under favorable circumstances, an acute inflammatory reaction is
followed within a few days by histological evidence that healing is
taking place. Large numbers of capillaries will be observed in the
granulation tissue of the wound area. The capillary branches grow
into the wound with its disorganized mass of fibrin, leukocytes and

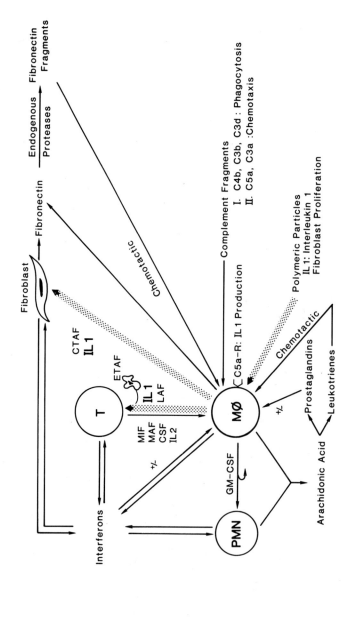

Fig. 6. The cellular and humoral regulatory pathways in the inflammatory response, mediated by the macrophage.

injured tissue. Macrophages present at the wound site release inter-
leukin 1 which stimulates the proliferation of fibroblasts. Fibroblasts are responsible for the synthesis of tropocollagen which subsequently aggregates to form collagen fibers, which are usually laid
down through a proteoglycan matrix. Over a period of time, the healing wound will show a decrease in leukocytes, capillaries, proteoglycans and water, and an increase in the amount of collagen (i.e.,
fibrosis). At still later time points the wound area will contain
mostly collagen with perhaps a few leukocytes. The wound is now in
the final stages of healing. The wound healing process with time
around the empty cage implants has been investigated in detail by
Hering and coworkers[11].

Chronic Inflammation and the Foreign Body Reaction

The most favorable result of an inflammatory response is the
complete return to the normal structure and function of the injured
tissue. Unfortunately, with biomaterial implants this rarely occurs
and the result most commonly seen is that of reorganization where repair is through fibrosis. In addition, the presence of the biomaterial will almost inevitably lead to a period of chronic inflammation,
the intensity of which will depend on the size of the area involved
and the consequences of the previous leukocyte/material interactions.
Figure 7 illustrates the possible responses following biomaterial
implantation.

A chronic inflammatory response implies the continued presence
of the injurious agent which may be represented even by a relatively
inert biomaterial. Other factors which can provoke chronic inflammation include: the persistent release of cytotoxic agents from the
implant, an implant which causes physical irritation to neighboring
tissues, extensive surgical injury, bacterial infection or host factors such as poor blood supply or nutrition. Chronic inflammation
is characterized by the predominance of mononuclear leukocytes (see
Figure 1), particularly macrophages. One may also observe signs of
acute inflammation (i.e., PMNs) at foci within a chronic inflammatory
lesion as well as foreign body giant cells.

If the biomaterial is either too large to be phagocytized or
quickly degraded, fibrous tissue begins to form about the mass of leukocytes and encapsulates the foreign body in a dense capsule of connective tissue. This sequence of events is known as the foreign body
reaction and tends to isolate the implant from the rest of the body.
A foreign body granuloma is composed of the foreign body surrounded
by giant cells, macrophages, lymphocytes and a few PMNs encapsulated
by fibrous tissue. Accumulation of blood pigments, lipids or calcium
salts are sometimes observed within the granuloma and is indicative
of a continuing chronic inflammatory reaction.

INFLAMMATION

INJURY (i.e., Implantation)

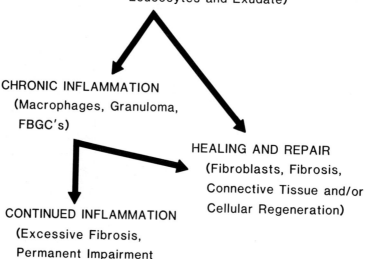

ACUTE INFLAMMATION

(Granulation Tissue, Proteins,

Leucocytes and Exudate)

CHRONIC INFLAMMATION
(Macrophages, Granuloma,
FBGC's)

HEALING AND REPAIR
(Fibroblasts, Fibrosis,
Connective Tissue and/or
Cellular Regeneration)

CONTINUED INFLAMMATION
(Excessive Fibrosis,
Permanent Impairment
of the Inflamed Area)

Fig. 7. The variations in the inflammatory response to an implanted material.

Multinucleated foreign body giant cells are commonly observed around biomaterial implants and are believed to be derived from the cytoplasmic fusion of macrophages after they have simultaneously attempted to phagocytize the same particle[12,13]. It appears inevitable therefore that giant cells will eventually form on the surface of a large biomaterial, providing that the macrophages are able to adhere to the surface. In our studies, utilizing the cage implant system in conjunction with light and scanning microscopy, we have observed giant cell formation on polymer surfaces as early as 4 days after implantation. Figure 8 shows a scanning electron micrograph of a giant cell along with other adherent macrophages on Biomer® 7 days after subcutaneous implantation. Figure 9 shows two well-developed giant cells on Biomer® after 21 days' implantation.

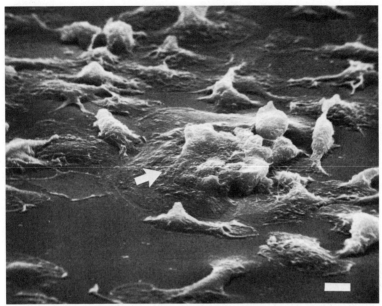

Fig. 8. Scanning electron micrograph. A foreign body giant cell
(arrow) and numerous macrophages adherent to a Biomer®
implant 7 days after implantation. The dimension bar
represents 10µm.

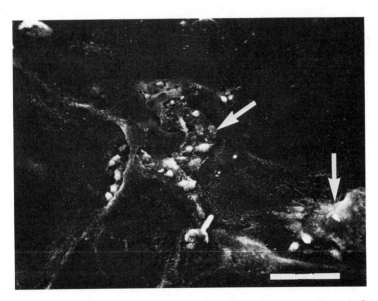

Fig. 9. Scanning electrom micrograph. Two well-developed foreign
body giant cells adherent to a Biomer® implant 21 days after
implantation. The dimension bar represents 100µm.

Histologically, there are two types of giant cells, the foreign body giant cell and the Langhan's giant cell, either of which may contain many nuclei. We have observed giant cells with greater than 50 nuclei on Biomer® surfaces by 21 days' implantation time[7]. The Langhan's type giant cells tend to have the nuclei located on the periphery of the cell and are commonly seen with infectious diseases such as tuberculosis, although they may also be seen in foreign body reactions. While the origin of foreign body giant cells appears to be well-established, the function of these cells, other than phagocytosis, remains unclear.

It should be apparent that both the physical and chemical properties of an implanted biomaterial will affect the intensity and duration of the foreign body reaction, including the concentration of leukocytes around the implant, the formation of giant cells and the thickness of the fibrous capsule. The attempt by the host to eliminate or neutralize the implanted material can also lead to eventual problems with implanted devices through fibrous encapsulation and calcification.

A comprehensive understanding of the cell and molecular interactions that occur with implanted biomaterials will provide considerable insight into the phenomenon which ultimately determines the biocompatibility of materials. The cage implant system is one approach which has enabled a more precise characterization of the crucial interactions.

REFERENCES

1. R.E. Marchant, A. Hiltner, C. Hamlin, A. Rabinovitch, R. Slobodkin and J.M. Anderson, In vivo biocompatibility. I. The cage implant system and a biodegradable hydrogel. J. Biomed. Mater. Res., 17:301 (1983).
2. J.M. Anderson, R.E. Marchant, S. Suzuki, K. Phua, C. Hamlin, A. Rabinovitch and A. Hiltner, In vivo biocompatibility studies. IV. Biomer® and the acute inflammatory response, in: "Polyurethanes in Medicine", H. Planck, editor, in press.
3. J.M. Anderson, R.E. Marchant and M. McClurken, Tissue response to drug delivery systems: the cage implant system, in: "Long-Acting Contraceptive Delivery Systems", G.I. Zatuchni et al., editors, Harper and Row, Inc., New York, in press.
4. C.W. Castor, Autocoid regulation of wound healing, in: "Tissue Repair and Regeneration", L.E. Glynn, editor, Elsevier/North-Holland Biomedical Press, New York (1981).
5. D.G. Wright and J.I. Gallin, Secretory response of human neutrophils: exocytosis of specific (secondary) granules by human neutrophils during adherence in vitro and during exudation in vivo. J. Immunol., 123:285 (1979).

6. P.M. Henson, Mechanisms of exocytosis in phagocytic inflammatory cells. Am. J. Pathol., 101:494 (1980).

7. R.E. Marchant and J.M. Anderson, unpublished results.

8. J.S. Sundomo and O. Götze, Human monocyte spreading induced by Factor Bb of the alternative pathway of complement activation. J. Exp. Med., 154:763 (1981).

9. I.M. Goldstein and H.D. Perez, Biologically active peptides derived from the fifth component of complement, in: "Progress in Hemostasis and Thrombosis", Volume 5, T.M. Spaet, editor, Grune and Stratton, New York (1980).

10. H.P. Hartung and U. Hadding, Complement components in relation to macrophage function. Agents and Actions, 13:415 (1983).

11. T.M. Hering, R.E. Marchant and J.M. Anderson, Type V collagen during granulation tissue development. Exp. Molec. Pathol., 39:219 (1983).

12. T.J. Chambers, Fusion of macrophages following simultaneous attempted phagocytosis of glutaraldehyde-fixed red cells. J. Pathol., 122:71 (1977).

13. A.R. Murch, M.D. Grounds, C.A. Marshall and J.M. Papadimitriou, Direct evidence that inflammatory multinucleated giant cells form by fusion. J. Pathol., 137:177 (1982).

MOLECULAR DESIGN OF MATERIALS HAVING AN ABILITY

TO DIFFERENTIATE LYMPHOCYTE SUBPOPULATIONS

Kazunori Kataoka

Department of Surgical Science, The Heart Institute of
Japan, Tokyo Women's Medical College
Shinjuku-ku, Tokyo, Japan

INTRODUCTION

Development of effective methods for cell separation is one
of the major objects in the field of biomedical science.[1] Separation of blood cells has become increasingly practical and more widely used in research and in the clinical field. Especially, there is
widespread utility of separating B and T cells, the two major subpopulations of lymphocytes. In the clinical field, their separation
is important in therapy and diagnosis of immunodiseases as well as
in assessing the matching of donor-recipient pairs in transplantation. In the field of fundamental and applied biology, preparative
separation of lymphocyte subpopulations is essential process for
producing high-value biological substances, including interferon,
lymphokine, and mono-clonal antibodies.

Adhesion chromatography, one of the promising methods for cell
separation, is a method which utilizes differences in the adhesive
properties of the cells with matrix surfaces as a basis for their
separation. Efficiency of adhesion chromatography is strongly
dependent on the ability of matrix surfaces to differentiate cell
populations in terms of adhesivity without undesirable perturbation
or activation of separating cells. To develop effective column
matrices, efforts should be made to clarify the effect of chemical
and physical properties of materials on the behavior of adhering
cells.

Although many studies have been carried out to describe the
behavior of cells at interfaces based on results obtained by macroscopic observations, including wettability measurements and electrophoresis, a generalized theory which clearly explains the behavior

225

of adhering cells has not been established yet. The plasma membrane of cells has a highly heterogeneous or mosaic structure composed of different types of molecules.[2] Changes in this mosaic structure caused by a change in molecular assemblage of these molecules will strongly affect the shape and function of adhering cells. This suggests that microscopic features of material surfaces should be taken into account for explaining the behavior of adhering cells and designing new matrices for adhesion chromatography.

In the course of our studies done to clarify the effect of surface charge of materials on cell adhesion, we have found that the shape and adhesivity of platelets were suppressed on microphase separated surfaces of polystyrene/polyamine comb-type copolymer (SA copolymer),[3,4] suggesting that undesirable contact activation of platelets was effectively suppressed on polymer surfaces having appropriate microdomain structures. To explain this unique behavior of SA copolymer, we have presented a hypothesis of "capping control" in which the microphase separated structure of the copolymer was assumed to regulate the shapes and adhesivity of platelets through its effect on the redistribution of membrane components (proteins and/or lipids) of the platelets.[5] Further, we have pointed out the feasibility of SA copolymer as a new column matrix for adhesion chromatography, because non-specific adhesion due to the contact activation could be effectively neglected on this copolymer surfaces.[3]

This paper describes the utility of microphase separated SA copolymer as a new column matrix for adhesion chromatography of B and T cells, and discusses the role of microphase separated structure in their differentiation.

EXPERIMENTAL

Materials. Preparation of polystyrene/polyamine comb-type copolymer (SA copolymer), polystyrene (PSt), poly(p-diethylaminoethylstyrene) (PEAS), and random copolymer of styrene and p-diethylaminoethylstyrene (P(St-EAS)) were reported elsewhere.[6] The structural formulas of these polymers are shown in Figure 1. A thin film casted on carbon-coated copper grid was stained by osmium tetroxide (OsO$_4$) vapor for 24 hrs. The micostructures of the film surface were observed by a transmission electron microscope.

Column preparation. Glass beads (40-60 mesh) were coated with one of the above-mentioned polymers by solvent evaporation technique under dry nitrogen atmosphere. A definite weight of polymer-coated beads was closely packed in the poly(vinyl chloride) tubing (3 mm ID) fitted with Nylon mesh column supports and a stopcock. The column primed with physiological saline was filled with 0.2 M phosphate buffer solution of rat serum albumin (0.09 g/dl), and was in-

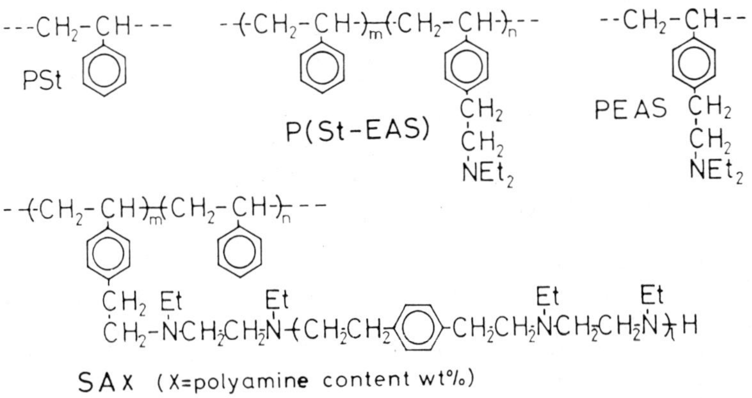

$---CH_2-CH---$ PSt

$-(CH_2-CH-)_m(CH_2-CH-)_n--$ P(St-EAS) CH_2 CH_2 $\overset{|}{N}Et_2$

$--CH_2-CH--$ PEAS CH_2 CH_2 $\overset{|}{N}Et_2$

$-(CH_2-CH)_m(CH_2-CH)_n--$

CH_2 Et   Et
$CH_2-\overset{|}{N}CH_2CH_2\overset{|}{N}(CH_2CH_2$⟨○⟩$CH_2CH_2\overset{|}{N}CH_2CH_2\overset{|}{N})_xH$

SAX (X=polyamine content wt%)

Figure 1 Structural Formulas of Polymers Examined

cubated for 12-18 hrs. The albumin solution in the column was replaced with Ca^{2+}- and Mg^{2+}- free Hanks' balanced salt solution
(HBSS) just prior to the lymphocyte loading.

Estimation of lymphocyte retention. Lymphocytes were obtained
from mesenteric lymphnode of Wister male rats aged 5 weeks, and suspended in HBSS in a concentration of $(1.5 \pm 0.1) \times 10^4$ cells/mm^3.
Purity of the lymphocytes was confirmed microscopically. The lymphocyte suspension was passed through the column using an infusion pump
for a definite period of time at a flow rate of 0.4 ml/min. Lymphocyte counts in the column-outflow were performed with a hemacyto-
meter. The beads situated in the upper part of the column were fixed
for 24 hrs in 1.25% glutaraldehyde. Thereafter, the specimens were
washed with water, freeze-dried, coated with gold, and examined in a
scanning electron microscope.

Identification of lymphocyte subpopulations. The percentage of
B cells, which express immunoglobulins (Ig) on their membrane surface, was determined by immunofluorescence staining of Ig molecules,
using an FITC-labeled rabbit anti-rat IgG.[6] The proportion of non-B
cells can be regarded as that of B cells, because lymphnode cell
populations are predominantly composed of B and T cells (>95%), with
negligible amount of null cells (<5%).

Estimation of albumin adsorption. ^{125}I-labeled human serum albumin was dissolved in 0.2 M phosphate buffer solution to form a 1/50
physiological concentration (0.09 g/dl). Polymer-coated beads were
incubated in the solution of ^{125}I-labeled albumin for varying periods
of time. Then, beads were rinsed repeatedly with 0.2 M phosphate
buffer using a vortex mixer. Each rinsing time was 10 seconds. Between each rinse, buffer was changed and the radioactivity of the
beads was counted by a γ-counter.

RESULTS AND DISCUSSION

Adhesion Behavior of Lymphnode Lymphocytes

As described in experimental part, effluent from the column was collected as a single aliquot. $([L]_0-[L])/[L]_0$ is defined as the retention of lymphocytes in the column, where $[L]_0$ is the concentration of lymphocytes prior to the loading in the column, and $[L]$ is the average concentration of lymphocytes eluted from the column. Table 1 shows the results of lymphocyte retention in the column packed with polymer-coated beads. These results were obtained for the columns of 10 cm long (weight of packed beads: 1g). As polystyrene/polyamine comb-type copolymer (SA copolymer) is a polystyrene derivative having long branches of polyamine, polystyrene (PSt) and poly(p-diethylaminoethylstyrene) (PEAS) can be regarded as models for trunks and branches of SA copolymer, respectively. As shown in Table 1, lymphocyte retention was significantly decreased for SA copolymers by the albumin coating. Although polyamine content of SA copolymers is in the range between PSt and PEAS, the retention of lymphocytes in albuminated columns was not correlated with polyamine content, and steeply dropped for SA copolymers having polyamine branches of 9 to 25 wt%.

As retention of lymphocytes in the column is considered to be a function of contact time of lymphocytes with polymer surfaces, retention behavior of lymphocytes at different contact times was examined by changing the column length. As shown in Figure 2, a semilogarithmic plot of the lymphocyte effusion from the column

Table 1   Retention of Lymphnode Lymphocytes in The Column Packed with Polymer-coated Glass Beads

Polymer	N Contents in The Polymer wt%	Retention of Lymphocytes (%)	
		Without Albumin[a]	With Albumin[a]
PSt	0	66.9 ± 4.3 (7)	53.1 ± 2.3 (36)
SA 9	1.0	---------------	9.5 ± 3.8 (5)
SA15	1.7	59.0 ± 6.2 (5)	19.3 ± 2.0 (36)
SA25	2.8	---------------	18.5 ± 1.9 (5)
SA50	5.6	66.9 ± 5.5 (5)	38.3 ± 3.0 (21)
PEAS	6.9	94.6 ± 1.3 (6)	92.8 ± 1.4 (19)

a) The mean ± S.E.M. The numbers of data points are shown in parentheses.

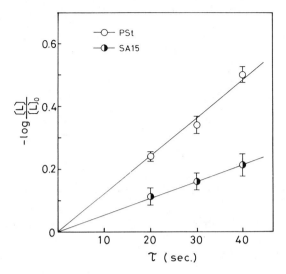

Figure 2   -Log($[L]/[L]_0$) vs. Time τ Plots for PSt and SA15 with
Albumin Precoating.   0.96 ml of lymphocyte suspension
was pumped through every column.

($[L]/[L]_0$) against the contact time was found to be linear.[6]   The
time (τ) required for the fluid to pass the column is plotted as
abscissa.   This time period was changed by changing the column
length.   The linearity in this semilogarithmic plot indicates that
the lymphocyte effusion can be empirically expressed by Eq. (1):

$$-Log([L]/[L]_0) = A_X\tau \text{ ----------------------------- (1)}$$

where A_X is a proportionality constant for material "X" and τ is the
time required for fluid to pass through the column.   The extent of
the interaction between lymphocytes and material "X" is considered
to be estimated by the value of a proportionality constant A_X.
Higher the value of A_X, higher the extent of the interaction of
lymphocytes with material "X".   Figure 2 clearly demonstrates that
lymphocytes interact more intensively with albumin-coated PSt than
with albumin-coated SA15.

To avoid the variation in the value of A_X caused by the effect
of individual differences in experimental animals, the value of a
proportionality constant for PSt (A_{PSt}) was adopted as an internal
standard, and the relative values of proportionality constants for
lymphocyte retention (A_X/A_{PSt}) were determined from the slopes of
straight lines obtained by plotting $-log([L]/[L]_0)_X$ against $-log($
$[L]/[L])_{PSt}$.   Results obtained for SA15 without adsorbed albumin
are shown in Figure 3.

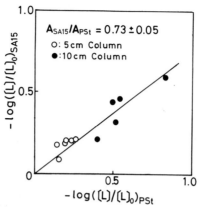

Figure 3 $-\text{Log}([L]/[L]_0)_{SA15}$ vs. $-\text{Log}([L]/[L]_0)_{PSt}$ Plots for Surfaces without Adsorbed Albumin

Table 2 summarizes the values of A_X/A_{PSt} ratios obtained for the various polymers examined.   Introduction of 9 wt% of polyamine branches in PSt backbone (SA9) brought about a steep decrease in the A_X/A_{PSt} ratio for the albumin-coated surface.   Such a steep decrease in the A_X/A_{PSt} ratio was not observed for the random copolymer of St and p-EAS (P(p-EAS)).   Although A_X/A_{PSt} ratios gradually increased with increasing the wt% of polyamine branches in SA copolymers from 9% to 50%, their values are still lower than that of PSt.   These results suggest that lymphocyte retention depends much more on the microphase separated structure rather than the amino group content of the polymers examined.

Figure 4 shows the transmission electron microscopic photographs of the microphase separated structure of the surface of SA copolymer films stained with vapor of osmium tetroxide (OsO_4).   Branches of polyamine, which can be stained with OsO_4, were separated into domains of circular islands in a sea of trunk polystyrene.   As the diameter of these polyamine islands was in the sub-micron range, each of the adhering lymphocytes must cover many microdomains of polyamine.

Figure 5 shows the typical scanning electron micrographs of adhering lymphocytes on albumin-coated surfaces.   Most of lymphocytes adhering to the surface of homopolymers (PSt and PEAS) suffered serious shape changes.   On the contrary, essentially all the lymphocyte population on SA copolymer surface retain there native spherical shape, and scarcely suffered any shape changes.   This suppressing

230

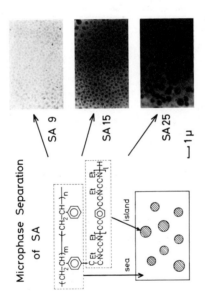

Figure 4   Microphase Separated Structures of SA Copolymer Surfaces Observed
By Transmission Electron Microscopy

Table 2 Values of A_X/A_{PSt} Ratios for Various Polymers

Polymer	N Contents in The Polymer wt%	Retention of Lymphocytes (%)	
		Without Albumin[a]	With Albumin[a]
PSt	0	1	1
P(St-EAS)	1.7	1.00 ± 0.03	0.70 ± 0.08
PEAS	6.9	1.98 ± 0.05	3.56 ± 0.03
SA 9	1.0	----------	0.14 ± 0.05
SA15	1.7	0.73 ± 0.05	0.18 ± 0.04
SA25	2.8	----------	0.24 ± 0.06
SA50	5.6	0.97 ± 0.03	0.72 ± 0.08

a) The mean ± S.E.M.

effect of SA copolymer on shape changes of adhering cells has been observed for platelets as well as lymphocytes,[4] suggesting that ability to suppress the shape change of adhering cells will be one of the most fundamental chracteristics of the microphase separated surface of SA copolymers.

Estimation of Selective Retention of Lymphocyte Subpopulations in The Column

Good linearity in a semilogarithmic plot of the lymphocyte effusion from the column against the contact time was also found to be valid for B cells and T cells, respectively. Results obtained for SA15 with adsorbed albumin are shown in Figure 6, indicating that preferential retention of B cells over T cells takes place on this graft copolymer surface. Thus, the effusion of B and T cells can be expressed as follows:

$$-\text{Log}([B]/[B]_0) = A_B \tau \quad \text{------------------------------} \quad (2)$$

$$-\text{Log}([T]/[T]_0) = A_T \tau \quad \text{------------------------------} \quad (3)$$

where [B]; concentration of eluted B cells, $[B]_0$; concentration of B cells prior to the loading in the column, A_B; a proportionality constant for B cell retention, [T]; concentration of eluted T cells, $[T]_0$; concentration of T cells prior to the loading in the column, A_T: a proportionality constant for T cell retention.

If the experimental conditions are constant, the value of the

a) PSt

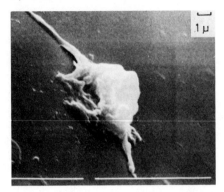

b) SA25

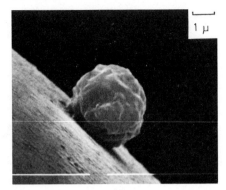

c) PEAS

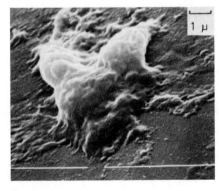

Figure 5   Scanning Electron Micrographs of Lymphocytes on albumin-coated surfaces of PSt, SA25, and PEAS

A_B/A_T ratio can be regarded as an inherent value for materials to be examined. Possible influence of the column length or the extent of retnetion will be excluded from the value obtained. We propose the value of A_B/A_T ratio as a measure for estimating selectivity for lymphocyte subpopulations. Namely, higher the value of this ratio, higher the selectivity for B cells. Values of the A_B/A_T ratios were determined from the slopes of straight lines obtained by plotting $-\log([B]/[B]_0)$ with $-\log([T]/[T]_0)$. Results obtained for PSt without albumin precoating are shown in Figure 7. Each experimental point represents the data obtained from a different rat, and nicely follows a straight line in a wide range. This linear relationship was confirmed to be valid for all of the materials examined.

The values of A_B/A_T ratios are summarized in Table 3. Albumin coating brought about the steep increase in the A_B/A_T ratios for SA copolymers, which forms a sharp contrast with a slight increase ob-

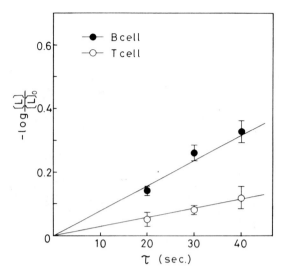

Figure 6   Logarithms of The Effusion Fraction of Lymphocyte vs.
Time τ Plots for B Cells and T Cells Eluted from An SA15
Column with Albumin Precoating.

served for PSt, P(St-EAS), and PEAS. Although comb-type copolymer,
SA15, and random copolymer, P(St-EAS), contain the same amount of
nitrogen (1.7 wt%) as amino groups, preferential retention of B cells
was observed only for SA15.

Figure 8 shows the relationship between the purity of eluted
T cells, $[T]/([T] + [B])$, and effusion fraction or yield of T cells,
$[T]/[T]_0$. These curves can be derived from the values of the A_B/A_T
ratios. By using the column packed with the beads coated with homopolymers (PSt and PEAS) or random copolymer (P(St-EAS)), the yield
of T cell populations with 95% purity is only 5%. On the other hand,
T cell population with 95% purity is obtained in 50% yield by using
the column packed with albuminated SA9 beads.

These results clearly indicate that adsorbed albumin layer
formed on the microphase separated surface of SA copolymer is essential for separating T cells from B cells. To characterize the
nature of adsorbed albumin layer formed on SA copolymer surfaces,
adsorption of albumin on polymer surfaces was then estimated.

Adsorption Behavior of Albumin on Polymer Surfaces

Figure 9 shows the results of time-varying studies of the adsorption of albumin on various polymer surfaces. Incubation time is
plotted as abscissa and adsorbed amount in $\mu g/cm^2$ as ordinate.
Dashed lines show the adsorption curves on homopolymers and random

234

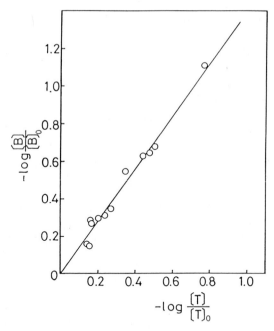

Figure 7   −Log([B]/[B]$_0$) vs. −Log([T]/[T]$_0$) Plots for PSt without
Albumin Precoating

Table 3   Values of A_B/A_T Ratios for Various Polymers

Polymer	N Contents in The Polymer wt%	A_B/A_T	
		Without Albumin[a]	With Albumin[a]
PSt	0	1.37 ± 0.02 (12)	1.56 ± 0.03 (27)
P(St−EAS)	1.7	1.43 ± 0.03 (12)	1.53 ± 0.04 (8)
PEAS	6.9	1.23 ± 0.04 (11)	1.30 ± 0.02 (17)
SA 9	1.0	----------------	4.47 ± 0.06 (18)
SA15	1.7	1.68 ± 0.06 (10)	2.89 ± 0.06 (28)
SA25	2.8	----------------	2.45 ± 0.06 (14)
SA50	5.6	1.42 ± 0.06 (12)	3.98 ± 0.03 (20)

a)   The mean ± S.E.M.   The numbers of data points are shown in
parentheses.

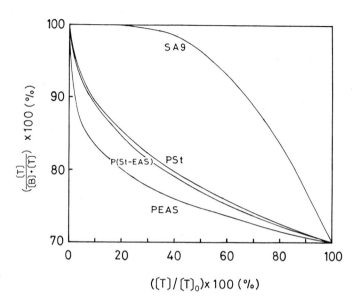

Figure 8   The Relationship between Purity of Eluted T Cells,
$[T]/([T] + [B])$, and Yield of T Cells, $[T]/[T]_0$. Percentage
of B cells in initial lymphocyte populations; 30%. Flow
rate; 0.4 ml/min. Volume of lymphocyte suspensions passing
through the column; 0.96 ml.

copolymer surfaces, and solid lines show that on SA copolymer surfaces. Adsorbed amounts reached plateau values after 8 hours of incubation for every surface. As shown in Figure 4, film surfaces of
SA copolymer were observed to be separated into two distinct microdomains of weakly cationic polyamine and neutral polystyrene. PSt
is regarded as the model surface of the polystyrene domain of SA copolymers and PEAS as the model surface of the polyamine domain of SA
copolymers. As can be seen from Figure 9, about 4 times more albumin
is adsorbed on PEAS than on PSt. This result demonstrates that adsorbed albumin on PSt is distinguishable from that on PEAS, in terms
of molecular packing or conformation.

There is another interesting phenomenon. Although amino group
content of SA50 is less than that of PEAS, more albumin was adsorbed
on SA50 than on PEAS. This curious phenomenon will be explained by
considering the desorption of adsorbed albumin during the rinsing
procedure. Figure 10 shows the change in adsorbed amounts of albumin
with the rinsing process. The amount of adsorbed albumin is plotted
as ordinate and the rinsing time as abscissa. On homopolymers and
random copolymer surfaces, amounts of adsorbed albumin decreased
steeply with rinsing. More than 70% of the initially adsorbed albumin was desorbed from the surfaces of PSt and PEAS. However. there
is only a 30-40% decrease for SA copolymer surfaces, indicating that

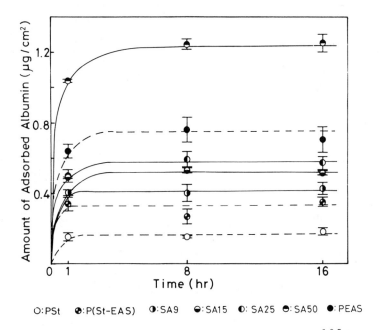

Figure 9 Time-varying Studies of The Adsorption of ^{125}I-Labeled Albumin on Polymer Surfaces

a more stable or tightly-adsorbed albumin layer was formed on SA copolymer surfaces compared with that on the homopolymer or random copolymer surfaces.

Based on these results, we consider that the albumin layer formed on the surface of SA copolymer has a heterogeneous mosaic structure which is organized corresponding to the pattern of the microphase separated structure of the SA copolymer itself. Therefore, adhering cells can recognize the microphase separated surfaces of SA copolymer through this organized protein layer, which may function as a messenger of information of the polymer surfaces.

Role of Microphase Separated Structures of SA Copolymers in Differential Retention of Lymphocyte Subpopulations

Lymphocytes attached on SA copolymer surfaces essentially retain their native spherical shape. This results suggest that contact activation of loaded lymphocytes would not occur in the albuminated SA copolymer columns. Thus, differentiation of B cell from T cell is probably not due to their differences in the degree of contact activation, but due to their differences in the physicochemical nature of plasma membrane surfaces. This is supported by our recent study on the effect of environment temperature on separability of SA copolymer columns.[7] Lymphocyte retention was effectively reduced by

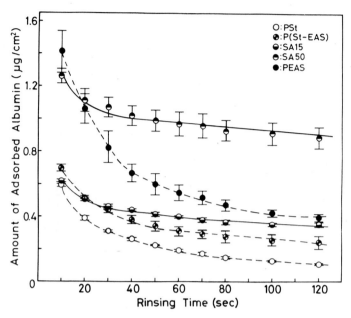

Figure 10   Change in Adsorbed Amounts of Albumin with Rinsing

introducing 9-25 wt% of polyamine branches in PSt backbone. This phenomenon is also explained by the suppression of active adhesion pathway of lymphocytes. Increase in lymphocyte retention with an increase in polyamine content of SA copolymers would be due to the enhancement of passive or physicochemical interaction between lymphocyte and polymer surface. Thus, even in SA50 column, B and T cells were effectively separated from each other, and retain spherical shape when attached on the beads surface.

We consider that stable microphase separated structure of adsorbed albumin layer on the SA copolymer play an essential role in suppressing the activation of lymphocytes. That is to say, microphase separated structure may suppress the activation of lymphocytes by regulating the excessive aggregation of plasma membrane proteins at cell-matrix interface (capping control mechanism).[4,5]

Practically, our results indicate that polymers having microphase separated structure will be a promising candidate as a column matrix for adhesion chromatography, because of their ability to suppress cellular activation processes.

ACKNOWLEDGEMENT

The author is grateful to Prof. Y. Sakurai and Dr. T. Okano,

Tokyo Women's Medical College, Prof. T. Tsuruta, Science University
of Tokyo, and Prof. S. Inoue, University of Tokyo, for their cooperation in carrying out the work cited in this article. The author
expresses his thanks to Dr. T. Nishimura, Mr. T. Watanabe, and
Mr. A. Maruyama for their continuous efforts to promote this research
project. This research was supported by the Ministry of Education,
Japan (Special Project Research, Design of Multiphase Biomedical
Materials) and by Japan Research Promotion Society for Cardiovascular Diseases.

REFERENCES

1. N. Catsimpoolas, ed., "Methods of Cell Separation, Vol. 1 & 2,"
    Plenum Press, New York (1977 & 1979).
2. G. L. Nicolson, Transmembrane Control of The Receptors on Normal
    and Tumor Cells, Biochim. Biophys. Acta 457:57 (1976).
3. K. Kataoka, T. Okano, Y. Sakurai, T. Nishimura, M. Maeda,
    S. Inoue, M. Shimada, I. Shinohara, T. Akaike, and T. Tsuruta,
    Development of New Biomaterials for Blood Cell Separator,
    Artificial Organs 5(Suppl):532 (1981).
4. K. Kataoka, T. Okano, Y. Sakurai, T. Nishimura, M. Maeda,
    S. Inoue, and T. Tsuruta, Effect of Microphase Separated
    Structure of Polystyrene/polyamine Graft Copolymer on Shapes
    of Adhering Rat Platelets in vitro, Biomaterials 3:237 (1982).
5. K. Kataoka, T. Okano, T. Akaike, Y. Sakurai, M. Maeda,
    T. Nishimura, Y. Nitadori, T. Tsuruta, M. Shimada, and
    I. Shinohara, Development of New Materials for Cell Separa-
    tion and Cell Culture: Estimation by Adhesion Chromatography,
    Jinko Zoki (Jap. J. Artif. Organs) 8:804 (1979).
6. K. Kataoka, T. Okano, Y. Sakurai, T. Nishimura, S. Inoue,
    T. Watanabe, A. Maruyama, and T. Tsuruta, Differential Reten-
    tion of Lymphocyte Subpopulations (B Cells and T Cells) on
    The Microphase Separated Surface of Polystyrene/polyamine
    Graft Copolymers, Eur. Polym. J. 19:979 (1983).
7. K. Kataoka, T. Okano, Y. Sakurai, T. Nishimura, S. Inoue,
    A. Maruyama, and T. Tsuruta, New Graft Copolymer of Polyamines
    for Chromatographic Separation of Lymphocyte Subpopulations,
    Artificial Organs, in press.

239

ATTACHMENT OF STAPHYLOCOCCI TO

VARIOUS SYNTHETIC POLYMERS

Anna Ludwicka[1], Bernd Jansen[2], Torkel Wadström[3],
Lech M. Switalski[3], Georg Peters[1] and Gerhard Pulverer[1]

[1]Institute of Hygiene, University of Cologne, Cologne
[2]Institute of Physical Chemistry, University of Cologne,
Cologne, Germany
[3]Department of Bacteriology and Epizootology, University
of Agricultural Sciences, Uppsala, Sweden

INTRODUCTION

Adhesion of cells to solid substrates has been the
subject of many investigations [1-6]. Most prior studies
on bioadhesion phenomena have dealt with the influence
of substrate surface properties on the relative adherence
and growth of mammalian cells [1,7]. Because of blood
compatability and infection problems the examination
of biomedical materials is of considerable clinical
interest. The attachment of bacteria to solid surfaces
like biomaterials is an important phenomenon because of
its possible role as the very first step in the development of an infection. Several parameters like critical
surface tension, surface free energy, charge and hydrophobicity, both of the substrate and the attached cells,
are surely involved in this phenomenon [1, 3-6, 8]. A relationship between the surface free energy of various
synthetic materials and the biological response (e.g.
cell adhesion or fibrous encapsulation) was postulated
[1, 6], but still the mechanisms of the interaction between bacterial cells and synthetic substrates is not yet
understood.

Staphylococcus epidermidis are the most common organisms
causing infections of indwelling artificial devices such
as ventriculo-atrial (-peritoneal) shunts, prosthetic
heart valves, joint prostheses, endocardial pacemaker
electrodes, continuous ambulatory peritoneal dialysis
(CAPD-) catheters and intravascular catheters [9-15].
Previous investigations have demonstrated that Staphylo-

241

coccus epidermidis are able to attach to and grow on surfaces of synthetic polymers and to produce an extra-cellular slime substance, which finally covers the staphylococcal cell layer [15-20]. In the following paper the attachment of Staphylococcus epidermidis strains to various synthetic polymers is investigated. The influence of surface properties of pure polymers and of surface-modified polymers on the attachment behaviour is examined.

MATERIALS AND METHODS

Bacterial strains and culture conditions

Five Staphylococcus epidermidis strains classified according to Kloos and Schleifer [21] from the culture collection of the Institute of Hygiene, University of Cologne, were studied: three of these strains (KH 6, KH 11 and KH 12) were isolated from infected intravenous catheters, the two other strains (Gloor 1/1 and Gloor 99) were cultivated from skin infections. The strains were cultured in nutrient broth (Difco) in Erlenmeyer flasks for 18 hrs. at 37°C. Bacteria were harvested by centrifugation (2500 x g, 15 min., 20°C) and washed three times in phosphate buffered saline (PBS, 0.13 M NaCL, 0.02 M phosphate buffer, pH 7.4). Cells were resuspended in PBS and the number of cells was measured using a Beckman spectrophotometer (A 600 nm) and compared with the results of colony counting on blood agar. The staphylococcal suspensions were then adjusted to 10^8 cells/ml.

Preparation of synthetic polymer surfaces

Synthetic polymers examined in our study are listed in Table 1. Small discs of the polymers (d= 6 mm) were cleaned with a detergent and ethanol followed by extensive rinsing in distilled water. The modified polymers were only washed in ethanol and extensively rinsed in distilled water. The dry samples were sealed in polyethylene bags and sterilized by irradiation (dose: 2.5 Mrad) using a ^{60}Co-γ-source (activity: 5000 Ci). Polymer pieces prepared in this way were used in all experiments.

Modification of surface charge and hydrophobicity of some synthetic polymers

Surface modified polyetherurethans were grafted with 2-hydroxyethylmethacrylate (HEMA) - a hydrophylic monomer - according to the method of Jansen and Ellinghorst[22] (see table 1). PVF$_2$ was modified by grafting with acrylic acid (AAc) and thereafter treated with 5 % aqueous KOH-solution.

Contact angle measurements

Contact angles on synthetic polymers and on bacteria were performed using a contact angle geniometer of Fa. Lorentzen and Wetts (Stockholm) combined with a special cell (Ramé-Hart, New Jersey). Triple distilled water (for polymers) and physiological saline solution (for bacteria) were used as contact angle liquids. The polymer samples were cleaned with soap solution, ethanol and tripledistilled water prior to the measurements. The angles of 10 sessile drops on each sample were measured and the mean average taken as a final result (standard deviation: $\pm$ 3°). Contact angles on bacteria were determined according to a method described by van Oss[23]. Flat layers of staphylococci were obtained by pouring washed bacterial cells (10^{10} cells/ml) on agar-covered microscopic slides, followed by air drying for at least 15 min.. Again, a set of 10 measurements was done with each sample (standard deviation: $\pm$ 2.5°).

Hydrophobic interaction chromatography (HIC)

Bacterial suspensions (100 µl, approx. 5 x 10^9 cells/ml) were applied on Octyl-Sepharose (Pharmacia) packed in Pasteur pipettes plugged with glass wool. Columns were equilibrated and eluted with 1 M ammonium sulfate in 0.1 M sodium phosphate pH 7.0. The absorbance (A_{600} nm) of 5 ml eluates was measured. The percentage of bacteria adsorbed to Octyl-Sepharose gel was calculated. The strains which adsorbed to the gel more than 75 % were considered as hydrophobic [24].

Measurements of bacterial attachment to synthetic polymers by a luminometric method.

250 µl of the staphylococcal suspensions (10^8 cells/ml) were given to the synthetic polymer pieces and the samples incubated at 20°C without agitation. After one hour incubation the polymer pieces were rinsed with 10 ml PBS to remove non-, or loosely adhered bacteria. The number of cells adhering to polymers was determined by the luminometric method using the LKB-Wallace-Luminometer[25]. After removing excessive PBS from polymers they were transfered to cuvettes containing 50 µl of 2.5 % trichloroacetic acid (TCA) and vigorously shaken

to extract ATP from attached cells. The amount of bacterial ATP in the extract was determined after addition of 1 ml of reconstituted ATP Monitoring Reagent (LKB - Wallace) with tris-EDTA buffer (0.1 M tris and 2 mM EDTA adjusted to pH 7.75 with a acetic acid), and calculated from ratio I/I_{stand}. multiplied by the amount of ATP in 20 μl of added ATP Standard (LKB-Wallace) (0.2 nmoles) [26, 27]. From this the blank (calculated in the same way) was subtracted to obtain the final result. The number of attached cells per cm² was calculated from standard curves prepared with known amounts of bacterial cells in suspension. Mean values of three samples of each polymer were taken as a final result (deviation less than 5 %).

RESULTS

Attachment of different S. epidermidis strains to various synthetic materials was investigated using the bioluminescence method [25].Fig. 1 shows the kinetics of staphylococcal strain attachment to polyethylene. Measurable attachment occurred within minutes, saturation was obtained after 30 - 60 minutes and extension of the incubation time beyond this resulted only in minor changes of the number of cells attached. A one hour incubation time was, therefore, chosen for all experiments. As shown in Fig. 1 the kinetics of attachment of staphylococcal cells were independent of the cell density of the suspensions (10^7 - 10^9 cells/ml).

Some surface properties of the polymers examined such as water contact angle, surface free energy γ_{sv} and polar contribution to surface free energy γ_s^p are presented in table 1 (either literature data or calculated from our measurements[22]). The polymers in the table are divided into three groups: the first one contains commercially available unmodified polymers used as basic materials for medical applications. The second group comprises two surface modified polyetherurethanes (grafted with HEMA, a hydrophilic monomer, according to the method described by Jansen and Ellinghorst [22])with increased surface hydrophilicity. In the third group surface modified PVF_2 (grafted with acrylic acid and treated with KOH) and glass are included; these last two substance groups exhibit an extreme hydrophilicity and have a negatively charged surface, whereas all other polymers are noncharged.

Attachment of the S. epidermidis strains KH 11, KH 12 and Gloor 1/1 to the neutral, unmodified polymers

244

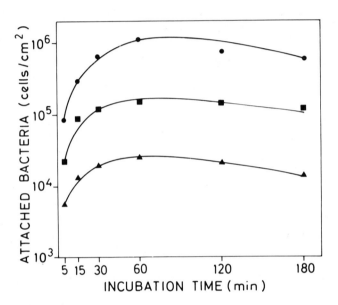

Figure 1   The kinetics of attachment of <u>S. epidermidis</u> KH 12 strain to
polyethylene at various cell densities: ▲10^7 cells/ml, ■ 10^8
cells/ml, ● 10^9 cells/ml.

TABLE I: SURFACE PROPERTIES OF THE POLYMERS TESTED

Polymer	Obtained From	Contact Angle (Water vs. Air) (°)	Surface Free Energy (γ_{sv}) (mN·m^{-1})	Polar Contribution To Surface Free Energy (γ_s^p) (mN·m^{-1})
Silicone Rubber (Silastic®)	Dow Corning Corp., Midland, USA	109	21[a]	0[a]
Polyetherurethane WH 8	Biosearch Inc., Raritan, N.J., USA	107	25[a]	2[a]
Polypropylene (amorphous)	Fa. Kalle, Wiesbaden, West Germany	104	30[a]	0[a]
Polyethylene	- " -	95	32[a]	0 - 2[a]
Polyethyleneterephtalate	- " -	78	40[a]	3[a]
Polyvinylidene-flouride (PVF$_2$)	Fa. Alkor, München, West Germany	77	33[a]	6[a]
Cellulose acetate	Fa. Lonza-Werke, Weil am Rhein, West Germany	58	45[a]	10[a]
Polyetherurethane WH 8 modified with HEMA; grafting yield 20%	Biosearch Inc., Raritan, N.J., USA	88	47[b]	20[b]
Polyetherurethane WH 8 modified with HEMA; grafting yield 130%	- " -	72	64[b]	30[b]
Polyvinylidene-fluoride grafted with AAC; grafting yield 20%; treated with 5% KOH-solution	Fa. Alkor, Munchen, West Germany	0	-	50[b]
Glass		0	170[a]	-

a = Literature data; b = calculated from our measurements[22]; c = modified by authors[22]

(with water contact angles ranging from 109° to 58°)
appears to be variable, but in most cases values of
more than 10^4 attached bacteria per cm² are found.
These polymers are either hydrophobic with no polar
groups or with moderate polar contributions to their
surface free energy. Attachment to modified polyetherurethans with higher values of polar contribution seems
to be less, and attachment to very hydrophilic and negatively charged modified PVF_2 and glass is generally
very low.

Fig. 2 shows the results for the strains KH 11, KH 12
and Gloor 1/1 plotted against the surface energy γ sv of
used solids. Although not all data fit in well, a curve
can be drawn showing a decrease of attachment with increasing γ sv and thus increasing hydrophilicity. To
emphasize this, the polar contribution to surface free
energy, $γ_{sv}^p$, of the solid substrate (either from literature data or from our measurements) was plotted
against the amount of attached bacteria (Fig. 3). The
majority of the points generally lies along a curve,
showing decreased attachment to those solids with high
ability for polar interactions.

The other two S. epidermidis strains KH 6 and Gloor 99
do not cause such great differences in attachment to
various materials, especially in the case of KH 6 strain
the bacterial amount is nearly on one level (exceptions
are modified PVF_2 and glass which show again very low
adhesion (Table 1). Determination of the hydrophobicity
of the bacterial strains (performed with contact angle
and HIC measurements) revealed that the strains KH 6
and Gloor 99 are more hydrophilic than strains KH 11,
KH 12 and Gloor 1/1 (Table 2). It can be seen that the
contact angle data are in good accordance with the results of the HIC measurements. In the case of the more
hydrophilic strains KH 6 and Gloor 99 the bacterial
attachment seems to be independent from γ sv of tne solid
substrate (Fig. 4).

DISCUSSION

There exist several types of interactions which par-
ticipate in the attachment process of cells to cellular
or solid substrates. Besides specific binding, where
adhesive parts on the surface of bacterial (adhesins)
as well on host cell membranes (receptors) are involved[28], nonspecific factors influencing the bacterial
adherence also occur[1]. These factors are important de-

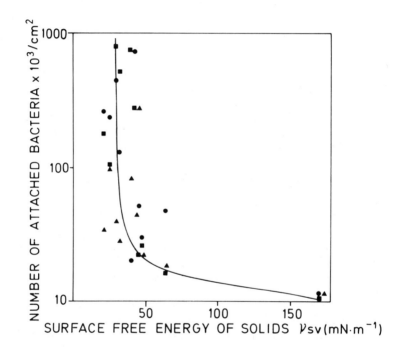

Figure 2 Attachment of three hydrophobic staphylococcal strains to several
solid substrates vs. surface free energy (γsv) of the
substrate ▲ S. epidermidis strain KH 11, ◼ S. epidermidis strain
KH 12, ● S. epidermidis strain Gloor 1/1. Number of bacteria in
the incubation mixture 10^8 cells/ml.

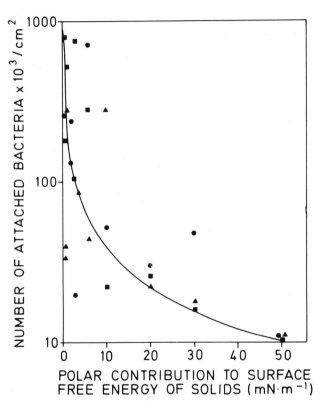

Figure 3 Attachment of three hydrophobic staphylococcal strains to various
solids vs. polar contribution to surface free energy (γ_s^p) of
the solids. ▲ S. epidermidis strain KH 11, ■ S. epidermidis strain
KH 12, ● S. epidermidis strain Gloor 1/1. Number of bacteria in
the incubation mixture 10^8 cells/ml.

TABLE II
Results of Contact Angle- and HIC-Measurements
for Several <u>Staphylococcus Epidermidis</u> Strains

Bacterial Strains	H2O – Contact Angle (°)	HIC-% of Adsorbed Bacteria
KH 11	46.5	> 90
KH 12	38.1	> 90
Gloor 1/1	46.0	> 90
KH 6	14.1	~ 20
Gloor 99	14.1	~ 20

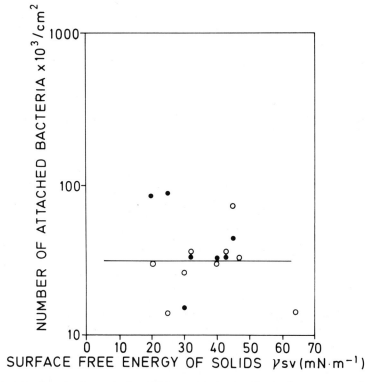

Figure 4   Attachment of two <u>hydrophilic staphylococcal strains to several polymeric materials</u> vs. the surface free energy (γsv) of the polymer o <u>S. epidermidis</u> KH 6, ● <u>S. epidermidis</u> Gloor 99. Number of bacteria in the incubation mixture 10^8 cells/ml.

terminants in the formation of multiple bonds between bacteria and host cells or solid substrates. Marshall et al.[29] reported that adhesion of marine bacterial specimens to solid, smooth surfaces is a spontaneous adsorptive phenomenon, where electrostatic forces and a dispersion balance between cells and the substrate in aqueous media is involved. These authors also suggested that the initial stage of bacterial attachment based on the theory of lyophobic colloid stability is reversible. A more permanent adhesion was observed when extracellular polymers interact between bacteria and the surface [29]. The net charge of the surface of both bacterial cells and host cells or solid surfaces may create repulsive forces between the cells or substrates[1,6]. Most of the reports concerning attachment to solid substrates deal with marine bacterial specimens[2, 3, 29], whereas very little is known about the attachment of staphylococci to synthetic polymers. This problem is of great importance because coagulase-negative staphylococci are very often involved in infections of prosthetic implants or medical devices[9-15], as was clearly demonstrated by us already in earlier publications [15-19]. Our investigations on the relationship between surface properties of unmodified and modified synthetic polymers and staphylococcal attachment shown in Figure 2 demonstrate that the adherence of hydrophobic strains of S. epidermidis is dependent on the surface free energy γ sv of the substrate. A decrease of attachment with increase of γ sv and thus increasing hydrophilicity was observed. Our results are in general accordance with the data of Fletcher[3], Hogt et al.[5] and of Absolom and Neumann[7], who developed a thermodynamic model for the adhesion of cells to solid surfaces. In their considerations, adhesion is governed by the change in free energy of adhesion (Δ F^{adh}) and is dependent on the surface free energy of the suspending liquid medium (γ LV) and of the solid substrate (γ SV) and of bacteria (γ BV). They distinguished between three possibilities:

1.) γ BV < γ LV: in this case Δ F^{adh} increases with increasing γ SV of the solid substrates, attachment should increase with increasing hydrophobicity (or decrease with increasing hydrophilicity),
2.) γ BV > γ LV: Δ F^{adh} decreases with increasing γSV and adhesion should decrease with increasing hydrophobicity,
3.) γ BV $\simeq$ γ LV: then Δ F^{adh} = 0, adhesion should be independent of the surface free energy of the solid substrate.

In our experiments γLV of the surrounding liquids (saline solution, PBS) with values of about 72 $mN.m^{-1}$ is greater than the surface free energy of the bacteria, which exhibit γ BV values of about 67 $mN.m^{-1}$ [30]. Therefore adhesion should decrease with increasing γ sv. As can be seen from Fig. 2 our results are in good agreement with the thermodynamic model.

A plot of our data against the water contact angle of synthetic polymers does not show such a clear relationship as found by other authors [3-5]. An explanation for this may be drawn from the findings of Holly [31], who stated that the water contact angle (measured in air) is strongly dependent on the orientation of polymer chains at the surface and is therefore not in any case a useful parameter for surface hydrophilicity. A more clear relationship was found when our data were plotted against polar contribution to surface free energy of polymers (Fig. 3).

The attachment of more hydrophilic strains KH 6 and Gloor 99 (see table 2) to all synthetic polymers (Fig.4) does not show such a relationship to surface free energy as the hydrophobic strains do. In this case the surface free energy γ BV of the bacteria is higher and may approach the value of γ LV of the surrounding liquid (saline solution or PBS). As mentioned above for such a case [1], adhesion should be independent from γ SV of the solid substrate, and this is nearly true for our results (Fig. 4). Bacterial adhesion to the very hydrophilic, KOH-treated, PVF_2-g-AAc and glass is very low in all experiments. A reason for this might be the fact that both solids have also a negative surface charge. As most of the bacteria have also a negative surface charge, a repulsion of the bacteria from the solid due to electrostatic forces must be assumed [32].

From the data presented here it can be concluded that the interaction of staphylococcal strains with synthetic polymers is a multifactorial event. Hydrophobic interaction is an important factor, as could be shown. However, also electrostatic forces seem to be involved in the mechanisms of attachment, which in our opinion cannot be regulated by a single factor. Besides the forces found in this study other factors, not yet identified, may be existing and interfering with the complex mechanisms of attachment.

REFERENCES

1. R.E. Baier, Comments on cell adhesion to biomaterial surfaces: Conflicts and concerns. J. Biomed. Mat. Res. 16: 173 (1982).

2. W.A. Corpe, Attachment of marine bacteria to solid surfaces, in: "Adhesion in biological systems", R.S. Manly, ed., Academic press, New York (1970).

3. M. Fletcher, and G.L. Loeb, Influence of substratum characteristics on the attachment of a marine pseudomonad to solid surfaces. Appl. Environ. Microbiol. 37: 67 (1979).

4. D.E. Gerson, and D. Scheer, Cell surface energy, contact angels and phase partition. III. Adhesion of bacterial cells to hydrophobic surfaces. Biochem. Biophys. Acta 602: 506 (1980).

5. A.H. Hogt, J. Feijen, J. Dankert, and J.A. de Vries, Adhesion of Staphylococcus epidermidis and Staphylococcus saprophyticus onto FEP-teflon and cellulose acetate. Abstract, International Conference on Biomedical Polymers, Durham, G.B. (1982).

6. D. Lerche, The adhesiveness to the cell surface with regard to it's electric and steric structure. Bioelectrochem. Bioenerg. 8: 293 (1981).

7. D.R. Absolom, A.W. Neumann, W. Zingg, and C.J. van Oss, Thermodynamic studies of cellular adhesion. Trans. Amer. Soc. Artif. Int. Organs XXV: 152 (1979).

8. M.E. Schrader, On adhesion of biological substances to low energy solid surfaces. J. Colloid. Interf. Sci. 88: 288 (1982).

9. R.J. Holt, Bacteriological studies on colonized ventriculoatrial shunts. Dev. Med. Child. Neurol. 12 (Suppl.22): 83 (1970).

10. H. Masur, and W.D. Johnson, Prosthetic valve endocarditis. J. Thor. Cardiovasc. Surg. 80: 31 (1980).

11. F.E. Stinchfield, L.U. Bigliani, H.C. Neu, Th.P.Gross, and C.R. Foster, Late hematogenous infection of total joint replacement. J. Bone Joint Surg. 62 A: 1345 (1980).

12. M.H. Choo, D.R. Holmes, B.J. Gersh, J.D. Maloney, J. Merideth, J.R. Pluth, and J. Trusty, Permanent pacemaker infections: Characterization and management. Am. J. Cardiol. 48: 559 (1981).

13. J. Rubin, W.A. Rogers, H.M. Taylor, E.D. Everett, B.F. Prowant, L.V. Fruto, and K.D. Nolf, Peritonitis during continuous ambulatory peritoneal dialysis. Ann. Int. Med. 92: 7 (1980).

14. G.L. Archer, Antimicrobial susceptibility and selection of resistance among Staphylococcus epidermidis isolates recovered from patients with infections of indwelling foreign devices. Antimicrobiol. Agents Chemo ther.14: 353 (1978).

15. G. Peters, G. Pulverer, and R. Locci, Bakteriell infizierte Venenkatheter. Dtsch. Med. Wochenschr. 106: 822 (1981).

16. G. Peters, R. Locci, and G. Pulverer, Microbial colonization of prosthetic devices. II. Scanning electron microscopy of naturally infected intravenous catheters. Zbl. Bakt. Hyg., I. Abt. Orig. B 173: 293 (1981).
17. R. Locci, G. Peters, and G. Pulverer, Microbial colonization of prosthetic devices. III. Adhesion of staphylococci to lumina of intravneous catheters perfused with bacterial suspensions. Zbl. Bakt. Hyg., I. Abt. Orig. B 173: 300 (1981).
18. G. Peters, R. Locci, and G. Pulverer, Adherence and growth of coagulase-negative staphylococci on surfaces of intravenous catheters. J. Infect. Dis. 146: 479 (1982).
19. G. Peters, F. Saborowski, R. Locci, and G. Pulverer, Investigations on staphylococcal infection of intravenous endocardial pacemaker-electrodes. Am. Heart J. (in press).
20. G.D. Christensen, W.A. Simpson, A.L. Bisno, E.H. Beachey, Adherence of slime-producing strains of Staphylococcus epidermidis to smooth surfaces. Infect. Immun. 37: 318 (1982).
21. W.B. Kloos and K.H.Schleifer, The genus Staphylococcus, in: "Prokaryotes", M.P. Starr, H. Stolp, H.G. Trüper, A. Balows, H.G. Schlegel, eds., Springer, Berlin(1982).
22. B. Jansen, and G. Ellinghorst, Radiation initiated grafting of hydrophilic and reactive monomers on polyurethane for biomedical application. Radiat. Phys. Chem. 18: 1195 (1981).
23. C.J. van Oss, and C.F. Gillman, Phagocytosis as a surface phenomenon. I. Contact angles and phagocytosis of non-opsonized bacteria. J. Reticuloendothel. Soc. 12: 283 (1972).
24. T. Wadström, S. Hjerten, P. Jonsson, and S. Tylewska, Hydrophobic surface properties of Staphylococcus aureus, Staphylococcus saprophyticus and Streptococcus pyogenes: A comparative study, in: Staphylococci and staphylococcal infections, J. Jeljaszewicz, ed., Gustav Fischer Verlag, Stuttgart-New York (1981).
25. A. Ludwicka, L.M. Switalski, A. Lundin, G. Pulverer, and T. Wadström, Measurement of bacterial attachment to a solid surface (synthetic polymers) by a bioluminescent assay. J. Envir. Microbiol., in press.
26. A. Lundin, Applications of firefly luciferase, in: Luminescent assay: Perspectives in endocrinology and clinical chemistry, M. Serio and M. Pazzagli, eds., Raven Press, New York (1982).
27. A. Lundin, and A. Thore, Comparison of method for extraction of bacterial adenine nucleotides determined by firefly assay. Appl. Microbiol. 30: 713 (1975).

28. E.H. Beachey, Bacterial adherence: Adhesin-receptor interactions mediating the attachment of bacteria to mucosal surfaces. J. Infect. Dis. 143: 325 (1981).
29. K.C. Marshall, R. Stout, and R. Mitchell, Mechanism of the initial events in the sorption of marine bacteria to surfaces. J. Gen. Microbiol. 68: 337 (1971).
30. C.J. van Oss, C.F. Gillman, and A.W. Neumann, "Phagocytic engulfment and cell adhesiveness as cellular surface phenomena", Marcel Dekker, New York (1975).
31. F.J. Holly, and M.F. Refojo, Wettability of Hydrogels I. Poly (2-Hydroxyethylmethacrylate). J. Biomed. Mater. Res. 9: 315 (1975).
32. V.P. Harden, and J.O. Harris, The isoelectric point of bacterial cells. J. Bacteriol. 65: 198 (1953).

BLOOD COMPATIBILITY OF POLYETHYLENE AND OXIDIZED POLYETHYLENE IN A

CANINE A-V SERIES SHUNT: RELATIONSHIP TO SURFACE PROPERTIES

Michael D. Lelah, Carol A. Jordan*, Mary E. Pariso,
Linda K. Lambrecht, Ralph M. Albrecht* and
Stuart L. Cooper

Department of Chemical Engineering
and Schools of Pharmacy and Medicine*
University of Wisconsin-Madison
Madison, Wisconsin

INTRODUCTION

The contact of blood with a polymer surface results in the initial deposition of proteins, platelets, and other formed elements. Proteins deposit during the first moments of blood contact[1], while platelets start to adhere after about one minute of blood contact, when the protein layer is about 200 Å thick[2]. The polymerization of fibrinogen to fibrin, and the activation and aggregation of platelets, lead to thrombus formation and growth on the artificial surface.

Blood-surface interactions depend greatly on the nature of the polymer surface[3,4,5]. The surface chemistry and morphology affects the composition and conformation of the initially deposited proteins. These initial events subsequently determine the magnitude and extent of platelet activation, aggregation, and thrombus formation.

In this study, the relationship between surface structure and thrombogenicity for a biomedical polymer, polyethylene, was investigated. Polyethylene is used in a number of blood-contacting applications, including chronic shunts and catheters[6]. The biomedical grade polyethylene was in the form of an extruded tube (PE). In addition, the same PE tubing was oxidized and etched with chromic acid (OX-PE). A multiprobe approach was used to characterize the surfaces of PE and OX-PE before blood contact in order to establish surface property information.

A recently developed canine ex-vivo femoral arteriovenous (A-V) series shunt technique[7,8] was used to study platelet deposition, and thrombus formation and embolization on PE and OX-PE. In this technique, radiolabeling techniques and scanning electron microscopy (SEM) were used to follow early (1/2 to 60 minutes) platelet and fibrinogen deposition, and platelet morphological changes on the two surfaces.

EXPERIMENTAL

Polymer Materials: 1/8" I.D. extruded Intramedic® PE-350 polyethylene (PE) was used as the polyethylene test material. Sections of this tubing were oxidized and etched with chromic acid (Chromerge®) to make oxidized polyethylene (OX-PE) according to the following procedure: (a) mild oxidation with Chromerge® at 60°C for 15 minutes (b) rinsing with excess dilute nitric acid to remove any inorganic residue (c) copious rinsing with distilled water (d) drying with nitrogen gas, followed by drying in a vacuum chamber. The PE and OX-PE tubings were rinsed with double distilled water prior to use.

Surface Characterization: Sections of the same tubing used in the blood contact experiments were used in the surface characterization experiments. Contact angle measurements were made using the captive bubble technique of Hamilton[9] with air and octane as the probe fluids. Five measurements were made for each probe fluid on each surface studied. The technique was modified to apply to curved specimens[5]. Electron spectroscopy for Chemical Analysis (ESCA) was performed using a Physical Electronics PhI 548 spectrometer using a 280W Mg anode at pass energies of 100 eV (broad scan) and 25 eV (high resolution scan). Attenuated total reflection infrared spectroscopy (ATR-IR) spectra were obtained using a Nicolet 7199 FTIR with a variable angle ATR accessory (Barnes 300) at a resolution of 2 cm^{-1}. Longitudinal slices of the polymers were placed with their inner surfaces in contact with 45° and 60° Germanium crystals. SEM analysis was performed using a JEOL 35C SEM at 7 kV accelerating voltage.

Animal selection: Adult mongrel dogs weighing 18-35 kg were selected by screening for platelet aggregability[10] to ADP and epinephrine, and for a normal range of platelet count (150,000-400,000/μl), fibrinogen level (100-300 mg/dl) and hematocrit (35-50%).

Animal Surgery: The surgical procedure has been described in detail[7,8]. Autologous platelets were labeled with ^{51}Cr[11] and injected into the dog 18 hours prior to surgery. Following anesthetization with thiamylal sodium, ^{125}I-labeled fibrinogen[12] was injected into the animal. The femoral artery and vein in one leg were exposed and ligated as shown in Figures 1 and 2. The artery and vein were then cannulated with the series shunt which consisted of a 7 cm long entrance region, followed by 5 cm test sections joined in series (Figure 2). The return section was 80 cm long. The shunt was initially filled with a degassed divalent cation-free Tyrodes solution to prevent blood-air contact. A small branch artery proximal to the shunt cannulation site was cannulated with an 18 gage polyethylene catheter attached to an I.V. extension set, which in turn was connected to a syringe containing Tyrodes solution. The purpose of the branch connection was to enable the blood to be flushed out of the shunt following a predetermined interval of blood flow through the shunt. The Tyrodes solution was introduced at a rate of about 1 ml/s (300 s^{-1} wall shear rate) to minimize shear stripping of adsorbed material.

Immediately following flushing, the joined test section (Figure 2) was removed and fixed with freshly-prepared 2% glutaraldehyde in 0.1 M phosphate buffer (pH 7.4) without surface-air contact. Each section was then subdivided under buffer. A 3.2 cm segment of each test surface was counted in an automatic gamma counter, while a 6 mm segment was examined by SEM.

A new set of identical test surfaces joined as a shunt section was inserted for each time period (1/2, 1, 2, 5, 10, 15, 20, 30, 45, and 60 minutes of blood contact). Each material was tested in at least three different animals. Polyvinylchloride and silicone rubber materials were tested in the same experiment[7]. This report, however, focuses on the comparison of the response on the PE and OX-PE surfaces. Each shunt section (for each time point, in each animal) contained two sections each of PE and OX-PE, in addition to two sections of polyvinylchloride and silicone rubber. The results for the latter two surfaces are discussed elsewhere[7]. Blood flow rates were measured by an electromagnetic flow transducer (Figures 1 and 2). Blood samples were removed at regular intervals throughout the experiment for radioactivity determinations, platelet count, fibrinogen level, hematocrit, activated thromboplastin time, thrombin time, plasma protamine paracoagulation and euglobulin lysis time. Platelet aggregability to ADP and epinephrine were monitored at the start and end of each experiment.

SEM Examination: Samples were fixed in 0.1 M phosphate buffered 2% gluteraldehyde for at least 24 hours. This was followed by serial dehydration for 10 minutes each in 15, 30, 50 and 70% ethanol/water, and 15 minutes each in 80, 85, 90, 95, and 100% ethanol/water. The samples were then dried by the critical point procedure using 100% ethanol as the intermediate fluid and liquid CO_2 as the transitional fluid[13]. Following mounting on aluminum stubs with silver conductive paint, the samples were sputter coated with 15 nm gold-palladium. A JEOL-JSM 35C scanning electron microscope was used to view the samples at 10 kV accelerating voltage.

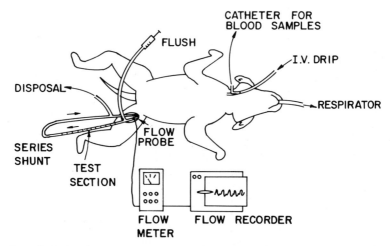

Figure 1. Schematic of animal experiment showing shunt, catheter, and flow measuring and recording instrumentation.

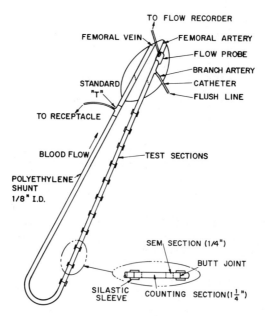

Figure 2. Cannulation site and shunt arrangement showing details of the 5 cm test sections connected in series.

RESULTS

Blood contact: Platelet deposition profiles as measured by radioactive labeling for PE and OX-PE are shown in Figure 3. The data
are the average results from three separate animal surgeries with
error bars representing standard deviations from the mean (where no
error bars are shown, the errors are within symbol dimensions).
For both surfaces there was an initial low rate of platelet deposition (from 1/2 to 5-10 minutes of blood contact) after which platelet deposition increased dramatically to a peak at 15-20 minutes of
blood contact. The peak deposition was higher on OX-PE (3090
platelets/1000 μm^2) and lower on PE (380 platelets/1000 μm^2).
Corresponding results for fibrinogen deposition are shown in Figure
4. Peak fibrinogen deposition was higher on OX-PE (2.82 $\mu g/cm^2$)
and lower on PE (0.29 $\mu g/cm^2$). The fibrinogen deposition profiles
show an initial decrease in fibrinogen deposition (from value at
1/2 min) to a minimum at 5-10 minutes of blood contact.

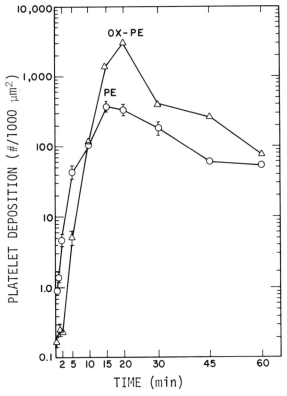

Figure 3. Platelet deposition profiles with time for PE and OX-PE.
Peak occurs at 15-20 minutes.

The sequence of scanning electron micrographs shown in Figure 5 shows the deposition and aggregation of platelets, thrombus formation and embolization processes occuring during blood contact on PE. At 1/2 minute of blood contact a few rounded platelets (some with short extensions) had deposited. Between 5 and 10 minutes a rapid increase in platelet deposition and aggregation was observed.

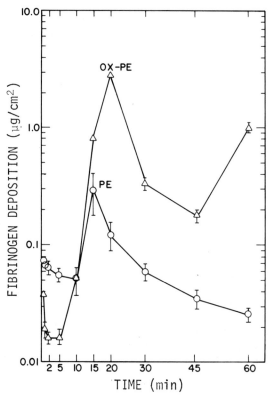

Figure 4. Fibrinogen deposition profiles with time for PE and OX-PE. Decrease from 1/2 min. to 5-10 min. of blood contact to peak at 15-20 minutes.

Between 10 and 20 minutes of blood contact, large platelet aggregates and thrombi were observed on the surface. A few white and red blood cells were also present. At 30 to 60 minutes, the thrombi appeared to have embolized from the surface, leaving single platelets or small platelet aggregates. The corresponding micrographs on OX-PE are shown in Figure 6. During the peak deposition period, large platelet thrombi were present with large numbers of red and white cells, and a network of fibrin associated with the thrombi (Figure 6b). The thrombi appeared to embolize from the surface between 30 and 60 minutes of blood contact (Figures 6c and d), leaving spread white cells with a few newly-deposited platelets.

Although thrombus formation was extensive on both PE and OX-PE, there were also separate regions of individual platelet deposition. A study of the shape change - activation of individual platelets in these regions was made. Figure 7 is a sequence of micrographs showing the shape change process occuring on individual platelets on the surface of PE. Between 5 and 15 minutes of blood contact there was substantial alteration of platelet morphology with pseudopod extension and degranulation and spreading of the hyalomere (Figures 7b and c). By 30 minutes of blood contact, individual platelets on the surface of polyethylene were spread almost completely over an area of about 25 μm^2. Figure 7d shows such spread platelets with recently deposited, more rounded platelets deposited on the spread platelets. The appearance of individual platelets on OX-PE (Figure 8) was similar to individual platelets observed on PE, except that the rate of shape change was more rapid on OX-PE, with spreading being completed by about 10 minutes of blood contact.

For all the experiments, mean blood flow rates were initially 350-400 ml/min. Flow rates decreased to 250-300 ml/min between 10 and 30 minutes of blood contact (the peak deposition period). By 45-60 minutes of blood contact, flow increased again to 300-400 ml/min.

Blood samples drawn at about 30-minute intervals during each experiment and at the end of each surgery indicated little variation in platelet count, fibrinogen level, activated partial thromboplastic time, prothrombin time, or euglobulin lysis time. Hematocrit decreased by about 10% and the PPP test turned positive towards the end of each experiment. Platelet aggregation to ADP and epinephrine decreased slightly by the end of each surgery.

Surface Characterization: Air-water and octane-water contact angles, measured using the captive bubble technique are shown in Table I. The air-water contact angle was higher (octane-water contact angle lower) on PE than on OX-PE indicating that the OX-PE surface was more hydrophilic than the PE surface.

Quantitative results from ESCA are shown in Table II. An oxygen peak was found both on PE and on OX-PE. A peak corresponding to silicon was measured on PE and on OX-PE. The O_{1S} high resolution ESCA peaks (Figure 9) show interesting results. The OX-PE oxygen peak is symmetrical except for a tail at lower binding energy. The PE oxygen peak has a secondary peak at lower binding energy, indicating the presence of at least two species of bound oxygen.

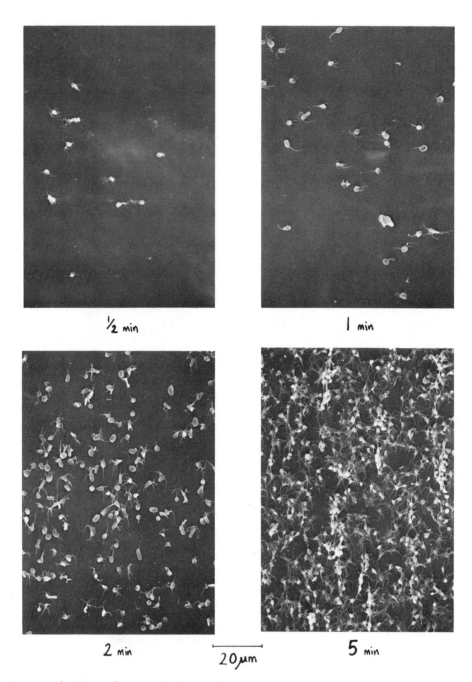

½ min   1 min

2 min   20μm   5 min

Figure 5(a,b,c,d): Scanning electron micrographs of the blood-exposed surfaces of PE at 1/2, 1, 2, and 5 minutes of blood contact. Note initial platelet deposition by 1/2 minute and rapid increase between 2 and 5 minutes.

264

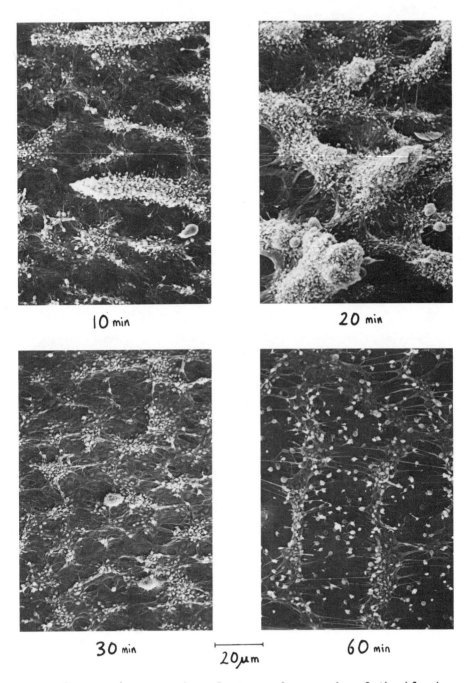

10 min   20 min

30 min   20μm   60 min

Figure 5(e,f,g,h): Scanning electron micrographs of the blood-exposed surfaces of PE at 10, 20, 30 and 60 minutes of blood contact. Thrombus formation occurs between 10 and 20 minutes, followed by embolization between 20 and 60 minutes.

265

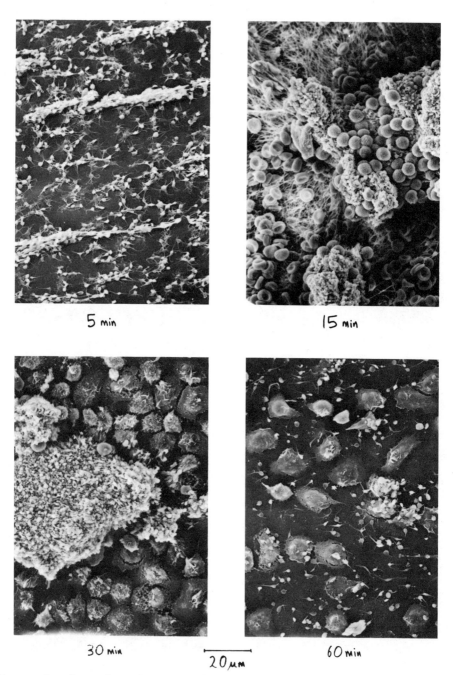

5 min   15 min

30 min   20μm   60 min

Figure 6: Scanning electron micrographs of the blood-exposed surfaces of OX-PE at 5, 15, 30 and 60 minutes of blood contact. Platelet aggregation is evident by 5 minutes. At the peak deposition (15 minutes) platelet thrombi have red and white cells associated. Embolization with white cell spreading occurs between 30 and 60 minutes.

266

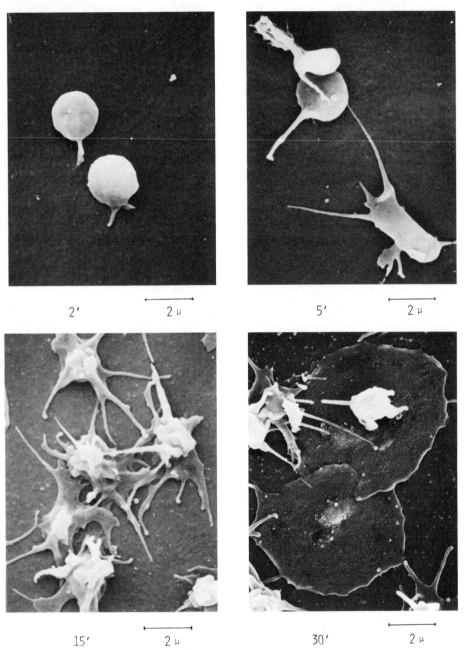

2'   2 μ   5'   2 μ

15'   2 μ   30'   2 μ

Figure 7:   Scanning electron micrographs of platelets not involved
in thrombus formation on PE at 2, 5, 15 and 30 minutes of blood
contact.   The sequence involves pseudopod extension (2-5 minutes),
degranulation (5-15 minutes) and spreading of the hyalomere (15-30
minutes).

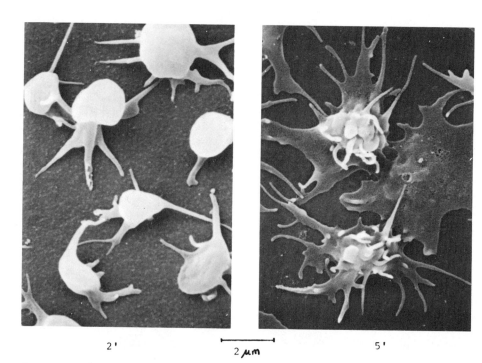

2'   2 μm   5'

Figure 8: Scanning electron micrographs of platelets not involved
in thrombus formation on OX-PE at 2 and 5 minutes of blood contact.
Marker is 2 μm. Platelet shape change is more rapid on OX-PE than
PE (compare Figure 7).

TABLE I

Contact Angle Data for PE and OX-PE

Surface	Air-Water Contact Angle	Octane-Water Contact Angle
PE	93°	spread
OX-PE	45°	78°

TABLE II

ESCA data for PE and OX-PE

Surface	Oxygen/Carbon	Silicon/Carbon
PE	0.072	0.030
OX-PE	0.22	0.049

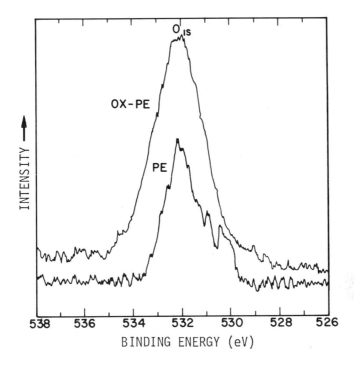

Figure 9: O_{1S} high resolution ESCA peaks for PE and OX-PE.

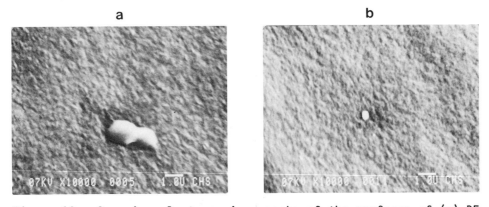

Figure 10: Scanning electron micrographs of the surfaces of (a) PE and (b) OX-PE at 45° tilt. Marker is 1 μm. The dust particles in the center of the micrographs were used for focusing.

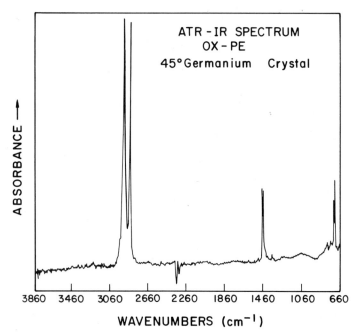

Figure 11: ATR-IR spectrum of OX-PE using 45° Germanium crystal.

TABLE III

Major IR peaks and assignments for PE and OX-PE

Wavenumber (cm^1)	Assignment
2917	Asymmetric C-H stretch
2848	Symmetric C-H stretch
1475	$-CH_2-$ bending
1460	(correlation splitting)
733	$-CH_2-$ rocking
722	(correlation splitting)

Figure 10 shows SEM photographs taken of PE and OX-PE before implantation, at 10,000 x magnification with a 45° tilt. The OX-PE surface does not appear to be etched or roughened, relative to the PE surface.

ATR-IR spectra were obtained from the surfaces of PE and OX-PE. Figure 11 shows the spectrum for OX-PE, which was very similar to that for PE (not shown). Table III lists the primary peaks and their assignments[14]. In both cases, no IR absorption peaks due to any oxygen-containing functional group were observed, using both 45° and 60° Germanium crystals. There were, however, small reproducible differences in the relative magnitudes of the split peaks. This information is presented in Table IV.

TABLE IV

Relative intensities of split peaks for PE and OX-PE

| Split peaks | Absorption Intensity Ratio | |
	PE	OX-PE
A_{2917}/A_{2848}	1.14	1.02
A_{1475}/A_{1460}	1.14	1.03
A_{733}/A_{722}	0.72	0.69

DISCUSSION

Blood compatibility: The acute canine ex-vivo femoral A-V series shunt technique used here allows for simultaneous monitoring of platelet and fibrinogen deposition and thromboembolism on a number of polymer surfaces using both SEM and radiolabeling techniques. The surgical technique allows for the measurement of the thromboembolytic potential of both polymers under similar ex-vivo physiological and hematological conditions, making a more precise comparison between polymers possible. The influence of the position and the nature (thrombogenicity) of any surface on the radiolabeling and SEM results of another section in the shunt has been previously investigated and found to be minimal under the experimental conditions used here[7]. Within the ±15% standard error determined for this technique, no effect of segment position has been observed.

The peak observed in both platelet and fibrinogen deposition (Figures 3 and 4), which is qualitatively confirmed by SEM (Figures

5 and 6), appears to be due to competition between thrombus formation and embolization. The presence of thrombi causes a reduction in the effective lumen size and therefore decreases blood flow rate, and potentially could lead to vessel occlusion. Embolization of large thrombi may result in tissue damage and infarcts downstream. The platelet peak height thus provides a measure of the acute blood compatibility of the surfaces. In this sense PE is more blood compatible than OX-PE. It should be noted however, that PE itself is a rather thrombogenic surface, in comparison with other polymers, such as silicone rubber, polyvinylchloride and polyurethanes[5].

SEM photographs (Figures 5 and 6) indicate that the morphology of the thrombi attached to PE is different from that observed for thrombi on OX-PE. Few white cells are associated with the thrombi on PE (Figure 5d) while large numbers of red cells, white cells and fibrin are associated with the thrombi on OX-PE. The large number of cells could be due to entrapment by the very large thrombi on OX-PE. Also, the white cells may adhere to the thrombi by chemotactic mechanisms due to the release of chemotactic agents (such as ADP and serotonin) by platelets fusing together during thrombus formation.

Observations of the activation-shape change of platelets not involved in the thrombus formation process (Figures 7 and 8), indicate that individual platelets on the protein covered surface undergo transformation at a much slower rate than when associated with thrombus formation and growth. In addition, these individual platelets undergo similar shape change activation, and over the same time period, as for single platelets observed in-vitro[15,16,17]. A number of conclusions may be drawn from these results. First, since the sequence of platelet activation-shape change is similar to that observed in-vitro[15,16] it appears therefore, that the activation sequence observed in-vitro is an intrinsic property of platelets, and not an artifact of the in-vitro preparation or test method. The morphological sequence thus does not depend on the presence of red or white cells, flow, or plasma proteins which may have been removed or added during in-vitro platelet preparation procedures.

Second, the rate of spreading is dependent on the nature of the surface. Platelets spread faster on OX-PE a more hydrophilic surface, than on PE, a more hydrophobic surface (Figures 7 and 8). These results are consistent with previous in-vitro work[15,16].

Third, the rate of spreading of individual platelets (Figures 7 and 8) is much slower than of those platelets involved in thrombus formation and growth (Figures 5 and 6). Thus, platelet activation can occur slowly, as in the case of individual platelets adhering to a surface or, the activation can be accelerated by or during platelet aggregation and thrombus formation, which is generally only observed in-vivo or ex-vivo. Baumgartner et al.[18] found that when platelet spreading did not occur significantly, platelets would saturate the surface at 10^4-10^5 platelets/mm^2 which is consistent with an individual platelet saturation value of about 40 platelets/1000 μm^2 (4×10^4 platelets/mm^2) observed in the present work. Baumgartner et al.[18] also found that if platelet spreading did occur, platelets would cover the surface. In the present work, rapid platelet spreading is associated with platelet aggregation and thrombus formation. It is speculated here that rapid platelet spreading may be initiated by the presence on the surface of a high concentration of a platelet adhesive protein such as fibronectin[19], the presence of leukocytes[20], or a combination of these or other factors.

In addition, there appears to be a relationship between the rate of individual platelet activation (Figures 7 and 8), and the overall platelet and fibrinogen response on PE and OX-PE. The greater rate of individual platelet activation on OX-PE correlates with higher peak platelet and fibrinogen levels, and thus poorer blood compatibility. This trend is also supported by other observations on polyvinylchloride and silicone rubber surfaces[21].

Surface Characterization: The OX-PE surface was more hydrophilic than the PE surface as determined by contact angle measurements (Table I). It is probable that the chromic acid treatment oxidizes surface oxygen containing species and hydrocarbon groups, increasing polarity. Rasmussen et al.[22] have investigated the chemical composition of chromic acid oxidized polyethylene. They found a surface containing about 60% carboxylic and 40% ketone or aldehyde moieties, which was highly hydrophilic.

The scanning electron micrographs (Figure 10) show that the surface of OX-PE is similar in texture to that of PE, at a 10,000 x magnification which is the scale of the size of a single platelet. This indicates that the etching procedure was quite mild. Visible etching has been observed under harsher conditions[22,23].

The ESCA analysis (Table II and Figure 9) provides elemental and bonding information to a depth of about 100 Å from the surface. The rather high surface oxygen level on PE may indicate surface oxidation during the extrusion process for fabrication of the tubing. The presence of silicon may be due to silicon-containing processing aids used in the extrusion. The high resolution O_{1S} spectrum for PE is of interest. The main peak at about 532 eV is probably due to carbon bonded to oxygen. The increase in the magnitude of this peak on OX-PE indicates further hydrocarbon oxidation. The secondary peaks on PE at a lower binding energy of about 530 eV are more difficult to interpret. It may be initially assumed that these peaks are due to oxygen-silicon or oxygen-carbon bonds (such as -Si-O-, -Si=O, C=O, C-O-C, COOH). However, these interactions would result in higher binding energies in contradiction to the lower binding energy peaks observed.

A comparison between the ESCA data and the ATR-IR spectral information is revealing. ATR-IR has a higher depth resolution (~1000 Å), and thus is less sensitive to surface features. No oxygen containing IR active group was found on either the PE or OX-PE (Figure 11), however oxygen was present on the surface as analyzed by ESCA (Table II). This indicates that the oxygen is highly localized on both the PE and OX-PE surfaces. The differences in intensities of split peaks (Table IV) may be related to differences in unit cell packing in the crystalline polyethylene due to attack by chromic acid. This effect may extend further from the surface into the bulk phase and thus is observable by ATR-IR.

Relationship between Blood Compatibility and Surface Chemistry: The OX-PE surface is more hydrophilic and more thrombogenic than the PE surface. Surface free energy values were not calculated as octane spread on PE. However, the results indicate that the PE surface has a lower surface energy than OX-PE. These implications for PE and OX-PE are in agreement with the results of Lyman et al.[24] and Ratner et al.[25]. Lyman et al.[24] have suggested that high surface energy correlates with increased platelet deposition. Ratner et al.[25] have found that platelet destruction was linearly related to the water content of hydrogels. Water content is a measure of hydrophilicity. Merrill et al.[26] however, have hypothesized that for polyurethanes, a hydrophilic surface deficient in ionic charge and hydrogen bonds is desirable for minimizing platelet retention and activation. The OX-PE surface, although hydrophilic, may be strongly ionic or have strong hydrogen bonding capability. These factors may override this opposing view on the relationship between blood compatibility and hydrophilicity. It appears that blood interaction is influenced by surface energetics (surface free energy, hydrophilicity, polarity etc.). However, the manner in which the surface energetics manifest themselves is not entirely clear.

The high resolution ESCA data showed that the increase in thrombogenicity observed in going from PE to OX-PE may be due to either the increase in the oxygen species exhibiting the main peak at 532 eV, or due to a decrease in the oxygen species giving rise to the secondary peak at 530 eV, or due to both. Unfortunately the nature of the latter species is not readily determined. Model compound studies are necessary to identify these species. The silicon/carbon ratio remained fairly constant before and after oxidation and thus probably does not play an important role in affecting thrombogenicity. Although the ATR-IR spectra for PE and OX-PE were almost identical, except for the differences noted in Table IV, the blood response was different. Use of a single surface probe such as ATR-IR would have thus led to false conclusions. This points to the desirability of a multi probe approach.

SUMMARY

An ex-vivo canine femoral A-V series shunt experiment was used to investigate the initial response of flowing nonanticoagulated blood to contact of PE and OX-PE. Platelet and fibrinogen deposition was greater on OX-PE than on PE as determined by radiolabeling techniques. Scanning electron microscopy was used to follow the formation, growth and embolization of thrombi, and was also used to follow the shape change activation of platelets not associated with thrombus formation. Thrombi on OX-PE were larger than on PE. Red and white cells, and fibrin networks were associated with the thrombi on OX-PE. Only occasional white cells were associated with the thrombi on PE. The sequence of pseudopod extension, degranulation, and spreading observed ex-vivo on individual platelets not associated with thrombus formation, was similar to that observed in-vitro. The rate of platelet shape change was greater on OX-PE, the more thrombogenic surface. Contact angle experiments showed that PE was more hydrophobic than OX-PE. ESCA results indicated the presence of oxidized species close to the surface of both PE and OX-PE. ATR-IR spectra showed that this oxidized layer was surface localized and did not extend significantly into the interior. The surface oxidized layer, which probably has a very different chemistry than the bulk polyethylene, appears to drastically affect the thrombogenicity of the polymer.

ACKNOWLEDGEMENT

The authors acknowledge the support of the National Institutes of Health, Heat, Lung and Blood Institute through grants HL-21001, HL-24046 and HL-29673. The hematological testing by Arlene P. Hart is gratefully acknowledged.

REFERENCES

1. R. E. Baier and R. C. Dutton, "Initial events in interactions of blood with a foreign surface", J. Biomed. Mat. Res., 3:191 (1969).
2. R. E. Baier, "Key events in blood interactions at nonphysiologic interfaces - a personal primer", Artific. Organs, 2 (1978).
3. A. S. Hoffman, "Blood-biomaterial interactions: An overview", in S. L. Cooper, N. A. Peppas (eds.) "Biomaterials: Interfacial Phenomena and Applications", ACS, 199:3 (1982).
4. J. D. Andrade, "Panel Conference: Blood-material interactions - 20 years of frustration", Trans. Am. Soc. Artif. Int. Org., 27:659 (1981).

5. M. D. Lelah, L. K. Lambrecht, B. R. Young, S. L. Cooper, "Physicochemical Characterization and in-vivo blood tolerability of cast and extruded Biomer", J. Biomed. Mater. Res., 17:1 (1983).

6. L. L. Hench, E. C. Ethridge, "Biomaterials - An Interfacial Approach", Academic Press, N.Y. (1982).

7. M. D. Lelah, L. K. Lambrecht, S. L. Cooper, "A canine ex-vivo series shunt for evaluating thrombus deposition on polymer surfaces", J. Biomed. Mat. Res. in press (1983).

8. M. D. Lelah, C. A. Jordan, M. E. Pariso, L. K. Lambrecht, S. L. Cooper, R. M. Albrecht, "Morphological changes occuring during thrombogenesis and embolization on bio-materials in a canine ex-vivo series shunt", SEM, in press (1983).

9. W. C. Hamilton, "A technique for the characterization of hydro-philic solid surfaces", J. Coll. Int. Sci., 10:219 (1972).

10. G. V. R. Born, "Aggregation of blood platelets by adenosine diphosphate and its reversal", Nature, 194:927 (1962).

11. A. F. Abrahamsen, "A modification of the technique of ^{51}Cr-labeled platelets giving increased circulating platelet radioactivity", Scand. J. Haematol., 5:53 (1968).

12. J. J. Marchalonis, "An enzymatic method for the trace iodina-tion of immunoglobulins and other proteins", Biochem. J., 113:299 (1969).

13. R. M. Albrecht, A. P. McKenzie, "Cultured and free living cells", in M. A. Hayat (ed.) "Principles and Techniques of Scanning Electron Microscopy", Vol. 3, p. 103, Van Nostrand Reinhold, N.Y. (1975).

14. P. C. Painter, M. M. Coleman, J. L. Koenig, "The theory of Vibrational Spectroscopy and its application to Polymeric Materials", John Wiley and Sons, N.Y. (1982).

15. R. D. Allen, L. R. Zacharski, S. T. Widirstky, R. Rosenstein, L. M. Zaitlin, D. R. Burgess, "Transformation and motility of human platelets: Details of the shape change and release reaction observed by optical and electron microscopy", Cell Biol. 83:126 (1979).

16. J. F. David-Ferriera, "The blood platelet: Electron microscopic studies", Int. Rev. Cytol., 17:99 (1964).

17. S. L. Goodman, M. D. Lelah, L. K. Lambrecht, S. L. Cooper, R. M. Albrecht, "In-vitro vs. ex-vivo platelet deposition on polymer surfaces", SEM, in press (1983).

18. H. R. Baumgartner, R. Muggli, T. B. Tschopp, V. T. Turitto, "Platelet adhesion, release and aggregation in flowing blood: Effects of surface properties and platelet function", Thrombos. Haemostas. (Stuttg.), 35:124 (1976).

19. B. R. Young, L. K. Lambrecht, S. L. Cooper, D. F. Mosher, "Plasma Proteins: Their role in initiating platelet and fibrin deposition on biomaterials", Adv. in Chem. Series ACS, 199:317 (1982).

20. T. A. Barber, T. Mathis, J. V. Ihlenfeld, S. L. Cooper, D. F. Mosher, "Short term interactions of blood with polymeric vascular graft materials: Protein adsorption, thrombus formation, and leukocyte deposition", SEM, II:431 (1978).
21. M. D. Lelah, unpublished results (1983).
22. J. R. Rasmussen, E. R. Stedronsky, G. M. Whitesides, "Introduction, modification and characterization of functional groups on the surface of low-density polyethylene film", J. Am. Chem. Soc., 99:4736 (1977).
23. J. S. Mijovic, J. A. Koutsky, "Etching of polymeric surfaces: A review", Polym.-Plast. Technol. Eng., 9:139 (1977).
24. D. J. Lyman, W. M. Muir, I. J. Lee, "The effect of chemical structure and surface properties of polymers on the coagulation of blood. I. Surface free energy effects", Trans. ASAIO, 11:301 (1965).
25. S. R. Hanson, L. A. Harker, B. D. Ratner, A. S. Hoffman, "Factors influencing platelet consumption by polyacrylamide hydrogels", Ann. Biomed. Eng., 7:357 (1979).
26. E. W. Merrill, E. W. Salzman, V. Sa Da Costa, D. Brier-Russell, A. Dincer, P. Pape, J. N. Lindon, "Platelet retention on polymer surfaces", Adv. Chem. Ser., 199:35 (1982).

POLYMER BASED DRUG DELIVERY:

MAGNETICALLY MODULATED AND BIOERODIBLE SYSTEMS

E.R. Edelman[ab], R.J. Linhardt[c], H. Bobeck,

J. Kost[bd], H.B. Rosen, and R. Langer[bde]

a   Medical Engineering and Medical Physics Program,
    Harvard-MIT, Division of Health Sciences and Technology
    Cambridge MA 02139, USA

b   Department of Surgical Research, Children's Hospital
    Medical Center, Boston MA 02115, USA

c   Division of Medicinal Chemistry, College of Pharmacy
    University of Iowa, Iowa City, IA 52242, USA

d   Department of Food and Nutrition, Massachusetts Institute
    of Technology, Cambridge MA 02139, USA

e   To whom correspondence should be adressed at MIT

INTRODUCTION

Polymer based drug delivery systems have been considered for
many applications to supplement standard means of medical
therapeutics. Currently nitroglycerin, scopolamine,
progesterone, and pilocarpine [1] are being administered on a
chronic basis from such devices. These delivery systems are less
complex than mechanical pumps and smaller because drug can be
stored as a dry powder within the polymer matrix. Recent
advances have shown that polymeric devices may be utilized for
very large molecular weight drugs [2], for drugs that must be
delivered in minute quantities [3], and at zero order kinetics
[4]. None-the-less two very important problems remain to be

279

answered. First, virtually all of these systems display rates of release that decay with time or are at the very best a constant function of time. None of them allows for control of drug release once the device has been implanted and release intiated. Second, after implantation and depletion of incorporated drug, these systems must be removed, as they are for the most part not biodegradable. Those polymers that have been proposed for biodegradable drug delivery systems generally progessively loosen because erosion is from the entire matrix bulk instead of just the surface. The result is that neither the rate of drug release nor polymer biodegradation is constant or predictable. This paper discusses ongoing research in our laboratory in these two areas.

REGULATION OF DRUG RELEASE FROM POROUS POLYMERS BY OSCILLATING MAGNETIC FIELDS

Ethylene-vinyl acetate copolymer (EVAc) is already used as the basis of polymeric delivery devices for low molecular weight drugs such as pilocarpine and progesterone. The material has been shown to be highly biocompatible [6] and suitable for use with macromolecules [2,5]. Recent investigators [7] have used them to formulate insulin delivery systems, and have controlled the blood sugars of diabetic rats for 120 days with an implant of 0.06 cc. Earlier studies showed that release rates from these systems could be regulated by the application of an oscillating magnetic field [8]. We now report on some of the important factors controlling modulated release of a macromolecule, bovine serum albumin (BSA), from such matrices.

Methods

The procedure for preparing the matrices was modified from earlier methods [8]. Polymer casting solution was made by dissolving EVAc in methylene chloride to achieve a 10% (w/w) solution. Powdered protein which had been seived to contain particles between 149 and 250 micrometers, was mixed with ten ml. of the casting solution. The suspension was poured onto a leveled glass mold which had been previously cooled by placing it on dry ice for five minutes. The mold remained on the dry ice throughout the procedure. Immediately following the pouring of the polymer-protein mixture, magnetic stainless steel spheres or beads were arrayed on the already cast layer using a specially constructed device. The beads were an alloy containing 79.17% iron, 17% chromium, 1% carbon, 1% manganese, 1% silicone, 0.75% molybdenum, 0.4% phospohrus, and 0.4% sulfur (Permag Northeast, Billerica, MA).

The loading device consisted of two rectangular plates of

plexiglass with identical arrangements of 131 holes, 1.8 mm in diameter spaced three mm apart. When the plates were shifted with respect to each other such that the upper and lower holes were offset, the magnetic beads rested in the holes of the top plate against the bottom. The plates could then be positioned over the mold and aligned such that the spheres fell in uniform array through the now aligned holes. Fifteen to 30 seconds after the magnetic beads were added, a top layer of polymer-protein mixture identical to the first layer was cast in the mold, as well. The mixture was left to solidify for ten minutes and then transferred to a -20°C freezer for 48 hours followed by drying at 20°C under houseline vacuum (600 millitorr) for an additional 24 hours. Final samples, one x one x 0.2 cm squares, were cut from the larger unit such that they contained nine magnetic spheres each. Paraffin (Fischer Scientific, paraplast) was melted then applied with a glass pipette to all but one face of the samples. Additional paraffin was used to mount each sample to the end of a polyvinylchloride rod (World Plastics, Waltham, MA) which passed through the cap of a twenty ml. glass scintillation vial. The depth of the rod within the vial was adjustable but was usually set so that the polymer slab face was one cm from the bottom of the vial. Ten ml. of 0.15 M sodium chloride or 0.1 M phosphate buffer was added and release media. Both solutions were made with double glass distilled water adjusted to pH 7.4 .

The oscillating magnetic field was generated from two five cm x five cm x 2.5 cm "Crucore-18" permanent magnets (Permag Northeast, Billerica, MA) mounted to a 1/4 inch thick plexiglass plate. This material is a complex alloy of the rare earth metal samarium cobalt which possesses a high residual induction, coercive force and maximum energy product. The magnetic moment was oriented though the top and bottom faces. The plate was attached to a motor and rotated beneath samples suspended in vials which were secured in holes in a second plexiglass plate.

BSA release from the polymer matrices was followed by absorbance at 220 or 280 nm. The media was changed before and after each field exposure and the duration noted. Rates of release were then determined by dividing the amount of protein released by this elapsed time. The ratio of the release rate under an applied field to the release rate when the field was withdrawn provided an assesment of the extent of modulation.

Results and Discussion

To assure that changes in absorbance of the release media were due to BSA alone, all of the materials involved in the experimental procedure was placed in the release media alone and then together, and then changes in absorbance determined. To determine whether the presence of the magnetic beads within the polymer matrix would alter normal release kinetics in the absence of an applied magnetic field, only half of the samples made from the same casting contained any beads. Release kinetics were followed for over a month and found to be identical ($p<0.0001$), and fit ($R^2 = 0.998$) linearly to the square root of cumulative time, suggesting that diffusion is the predominant release determinant. Finally, to show that modulation could only be achieved when there was both an applied alternating field and embedded magnetic particles, kinetic analysis was performed in the following experiments. 1) Slabs without any beads or embedded with 1.4 mm nonmagnetic stainless steel beads were exposed to magnetic fields alternating at frequencies ranging from 0.866 to 11 Hz, and 2) slabs with magnetic beads were examined in the absence of applied fields or in the presence of DC fields [9]. None of these groups exhibited any significant modulation.

Several sets of experiments were conducted to assess the affect of different system parameters on release. In one set, release was studied from identical slabs exposed to magnetic fields at different frequencies. The frequency was altered by increasing the rotational speed of the disc on which the permanent magnets were mounted. Frequencies of either 155, 200, 290 or 325 rotations per minute were used representing magnetic fields of 5.0, 6.6, 9.5, and 11.0 Hz. The frequency strongly affected the rate of protein release. After 180 hours of release at these four frequencies increases over baseline of 133%, 150%, 450% and 733% were obtained. The average ratio of the release rates obtained in the presence and absence of applied magnetic fields reveals a linear relationship with frequency ($R^2 = 0.996$), (figure 1). The magnetic particles probably move within the elastic environment of the EVA copolymer matrix. At very low frequencies the field is not alternating fast enough to create the optimum motion and alteration of the matrix. As the oscillation frequency increases so does the frequency of motion of the beads. At some critical frequency the visco-elastic properties of the copolymer material will probably no longer allow the bead full freedom of movement. We are currently investigating modulated release in the higher frequency ranges.

The force on an object in a magnetic field is proportional to the magnetic moment of the object and the strength of the magnetic field or its spatial gradient. Therefore, we expected this system to be sensitive to changes in these parameters. When the strength of the magnetic field was decreased 43.5%, from 2000 Gauss to 870 Gauss, by increasing the distance between the field source and the samples, the release rates during the field exposed periods fell 35% . When the amplitude was returned to its original strength the release rates returned, as well. Likewise, when a single 1.4 mm high, 1.4 mm diameter samarium cobalt permanently magnetized cylinder replaced the nine magnetic beads there was a sigificant difference in the release rates obtained in the presence and absence of magnetic fields (p<0.0001) (figure 2). The average ratio of the rates of release for the field exposure and field absent periods was 1.55 $\pm$ 0.35 for the beads and 12.4 $\pm$ 2.5 for the magnetized cylinders. The dependence on frequency and strength of the applied field provide two separate means of controlling the release rates from a device once implanted. In this manner, the same device could be used to deliver different amounts of drug in varying fashions, as required.

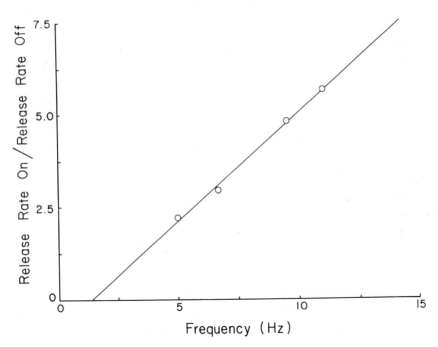

Figure 1: The average ratio of the rates of bovine serum albumin release from EVAc matrices during field applied and field absent periods plotted versus the frequency of magnetic field oscillation reveals a linear fit (R^2 = 0.996) with a slope of 0.59 . At 5.0 Hz the average ratio is 2.19 $\pm$ 0.46, at 6.67 Hz, 2.93 $\pm$ 1.0, at 9.5 Hz, 4.77 $\pm$ 0.63 and at 11.0 Hz, 5.62 $\pm$ 1.25 .

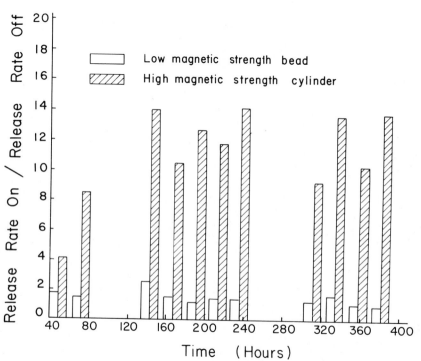

Figure 2: Ratio of rates of BSA release (averaged over all samples at a given time period) from EVAc matrices embedded with magnetic spheres (clear bars) and magnetized cylinders (hatched bars). Over the course of the entire experiment the average of this ratio was, 1.55 ± 0.35 for the bead embedded matrices and 12.4 ± 2.5 for those containing magnetized cylinders.

BIODEGRADABLE POLYANHYDRIDES

Although controlled release of drugs can be accomplished by several mechanisms [1], biodegradation of an insoluble polymer carrier to soluble monomer units offers the advantage of eliminating the need for surgical removal of the device. Controlled release matrices composed of hydrophilic biodegradable polymers such as poly(lactic acid), poly(glycolic acid), and their copolymers generally erode in a homogenous manner from the entire matrix and not just from the matrix surface [10,11]. This leads to a progressive loosening of the matrix which causes changes in both the permeability and mechanical strength of the devices during bioerosion [10,12]. It would be far more desirable if a matrix were to erode heterogeneously, from the surface first. Such erosion will lead to zero-order drug release provided that diffusional release is minimal and the overall shape of the device remains nearly constant; thus maintaining constant surface area [13]. To obtain a device that erodes heterogeneously, the polymer used should be hydrophobic yet contain water labile linkage. The only polymers designed for this purpose have been poly(orthoesters) [14]. However, because of the stability of the backbone bonds, these polymers erode slowly and require additives to promote biodegradation. The delivery systems containing additives, such as water soluble salts, swell considerably leading to diffusional release. It is possible that poly(anhydrides), which were originally synthesized as fiber forming polymers in the textile industry [15], but rejected because of their hydrolytic instability compared to polyesters of similar structure [15], might be sufficiently hydrolytically labile to produce heterogeneous erosion at rates suitable for controlled release applications.

Methods

Poly[bis(p-carboxyphenoxy) methane] (PCPM), was selected as a prototype poly(anhydride) because its erosion rate in NaOH [16] suggested to us sufficient hydrolytic labilty for drug delivery application. PCPM was synthesized by the method of Conix [16] from the mixed anhydride of bis(p-carboxyphenoxy)methane and acetic acid. The polycondensation was performed in vacuo at 195-210°C [16,17]. The Tg and Tm of the melt pressed polymer were determined on a Perkin-Elmer DSC-2 differential scanning colorimeter.

To formulate drug-free matrices, PCPM was ground using a Fisher Scientific Micro Mill and the resulting particles sized using sieves. PCPM particles (150-300 micrometer diameter) were melt-pressed between sheets of aluminum foil at 121.1°C and 8000 lbs. (representing from 22-81 kpsi) on a Carver Laborotory Press using shims to regulate device thickness. The matrix dimensions examined ranged from having face areas of 0.2 to 0.8 cm^2 and having thicknesses of 0.05 to 0.10 cm.

To incorporate a drug, unlabelled cholic acid from Sigma Chemical Co. (15mg) was dissolved in ten ml of ethanol. Titrated cholic acid from New England Nuclear, in 125 microliters of ethanol (2.42×10^8 DPM, S.A. = 3.72×10^{10} DPM/mg) was added to the unlabeled solution and the solution was dried in vacuo. The drug powder (10.5 wt %) was mixed with the PCPM particles and melt-pressed as above. The drug containing device was then sandwiched between two very thin (<1 wt %) drug free layers of ground PCPM to eliminate the presence of surface exposed drug particles and then repressed.

Erosion and drug release studies were performed by placing the PCPM devices weighing from 10-50 mg in glass scintillation vials containing 10 ml of 0.2 M sodium phosphate buffer (pH 7.4) at 37°C or 60°C. The buffer was periodically changed by removing the device from the vial and placing it in a vial of fresh buffer. The absorbance of the collected buffer solutions was measured on a Gilford spectrophotometer at 243 nm to detect the diacid monomer, bis(p-carboxyphenoxy) methane. Concentrations were determined from a standard curve constructed by measuring the absorbance at 243 nm of the pure diacid monomer at concentrations from 0 to 0.02 mg/ml. The polymer devices containing titrated cholic acid were eroded and the buffer solutions were analyzed on a Beckman scintillation counter.

Results and Discussion

The erosion curves for drug-free PCPM slabs at 37°C and 60°C are shown in figure 3. Both curves are characterized by an induction period followed by a linear region of mass loss at a nearly constant rate. Throughout the erosion, the devices decreased in size while maintaining their structural integrity [17] suggesting that only surface erosion is occurring.

Scraping the face of the polymer matrix did not eliminate the undesired induction period. However, when the samples were pre-eroded for 50 hours at 60°C, removed, vacuum dried, and returned to fresh buffer solutions at either 37°C or 60°C, near zero-order erosion of the matrices began almost immediately (figure 4). This suggests that two rate constants might control the rate of device erosion. The first is the rate of hydrolysis of the anhydride linkage and the second the rate of polymer dissolution. In order to begin to measure monomer units in the buffer it seems likely that many anhydride linkages on the polymer's surface must first be cleaved. This non-productive hydrolysis corresponds to the observed induction time. Pre-erosion of the device presumably decreases the surface polymer's chain length and in the subsequent erosion the anhydride hydrolysis rate and device dissolution rate become equivalent. Alternatively the observed induction period and its elimination by pre-erosion may be the result of an initially hydrophobic surface becoming increasingly hydrophilic as hydrolysis occurs. The rate of erosion would increase up to the point were it becomes limited by both the rate of hydrolysis and the rate of water penetration into the polymer.

The decrease in device thickness throughout the erosion [17] and the maintenance of the matrice's structural integrity, as well as, the nearly zero-order erosion kinetics suggests that the heterogeneous surface erosion predominates.

Drug release from PCPM was investigated using cholic acid which because of its low u.v. absorption at 234 nm did not interfere with the matrix erosion measurement. The erosion and release profiles were nearly zero-order, had similar slopes and both the drug and the polymer had completely disappeared at nearly identical times (figure 5); though initially, polymer erosion lagged behind drug release. This might be explained by one of two possible mechanisms. First, cholic acid which adheres to the surface of the polymer sample may be released upon contact with the buffered media, even before erosion occurs. Second, when a portion of the polymer slab does erode, cholic acid escapes and is immediately dissolved in the buffer solution. While the polymer material which breaks needs time to completely hydrolyze and dissolve to a point where it can be detected by our assay [17].

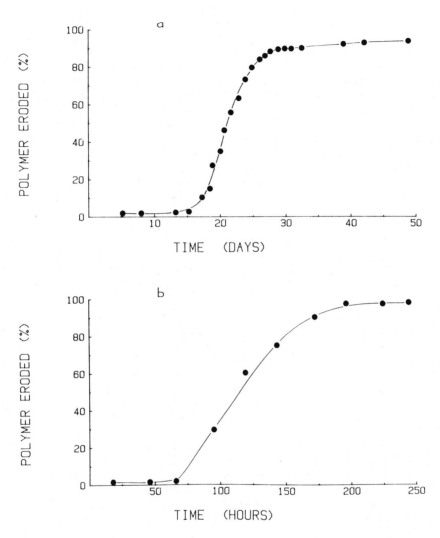

Figure 3:   Erosion curves for the drug-free PCPM matrices in phosphate buffer at (a) 37°C and (b) 60°C.   Percent polymer eroded is 100x the cumulative mass eroded at each sample point divided by the total mass of the matrices eroded.   The PCPM matrices eroded at 37°C and 60°C weighed 23 and 18 mg and had dimensions of 0.24 cm^2 face area x 0.08 cm thick and 0.33 cm^2 face area x 0.05 cm thick, respectively.

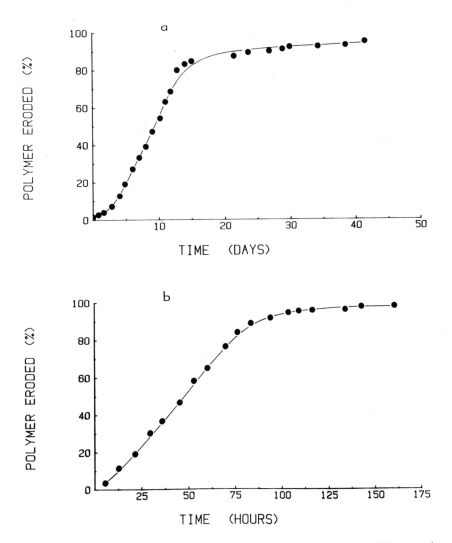

Figure 4:   Erosion curves for the for the drug-free PCPM matrices
which had been pre-eroded at 60°C for 50h, at (a) 37°C and (b)
60°C.   The matrices weighed 74 and 25 mg and had dimensions of
0.57 cm^2 face area x 0.06 cm thick and 1.13 cm^2 face area x 0.05
cm thick, respectively.

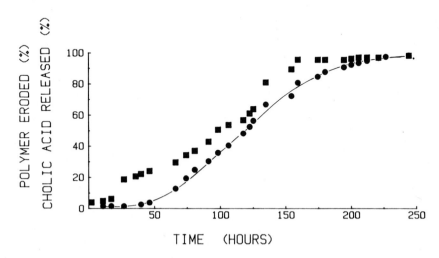

Figure 5: Erosion and release curves for a 25 mg PCPM matrix containing cholic acid at 10.5 wt.%, 0.5 cm^2 face area x 0.006 cm thick, eroded in phosphate buffer at 60°C. PCPM erosion (circles) is plotted as percent polymer erosion which equals 100x the cumulative mass eroded at each sample point divided by the total mass of the matrix. Cholic acid release (squares) is plotted as the percent cholic acid released which equals 100x the cumulative counts per minute at each sample point divided by the total number of counts per minute within the polymer matrix.

CONCLUSION

Reproducible regulation of macromolecule (bovine serum albumin) release from biocompatible polymer systems has been demonstrated. Small magnetic particles were embedded within the polymer matrix which was then subjected to an oscillating magnetic field. Parameters critical to the regulation of this release included the pole strength of the embedded particles and the amplitude and frequency of the applied magnetic field. A second polymer system, based on bioerodible polyanhydrides, displayed heterogeneous erosion and near zero-order release of a model drug (cholic acid).

ACKNOWLEDGEMENTS

The authors would like to thank Pamela Brown for assistance in preparing this manuscript and Dr. Larry Brown for his helpful comments in reviewing this paper. These studies were supported by N.I.H. grant no. GM 26698 and a grant from the Juvenile Diabetes Foundation.

REFERENCES

1 R. Langer and N.A. Peppas, Present and Future Applications of Biomaterials in Controlled Drug Release, Biomaterials , 2:201-214 (1981)

2 W. Rhine, D. Hsieh, and R. Langer, Polymers for Sustained Macromolecule Release: Procedures to Fabricate Reproducible Delivery Systems and Control Release Kinetics. J. Pharm. Sci., 69:265-270 (1980)

3 J. Murray, L. Brown, M. Klagsburn and R. Langer, A Micro System for Sustained Release of Epidermal Growth Factor, In Vitro, 19:743-750, (1983)

4 D. Hsieh, W. Rhine and R. Langer, Zero-Order Controlled Release Polymer Matrices for Micro- and Macromolecules, J. Pharm. Sci., 72:17-22 (1983)

5 R. Langer and J. Folkman, Polymers for the Sustained Release of Macromolecules and Other Proteins, NATURE, 263:797 (1976)

6 R. Langer, H. Brem, and D. Tapper, Biocompatability of Polymeric Drug Delivery Systems for Macromolecules, J. Biomed. Mat. Res., 15:267-277 (1981)

7   L. Brown, C. Munoz, L. Siemer, E. Edelman, J. Kost, and R. Langer, Sustained Insulin Release from Implantable Polymers Diabetes 32, Supplement #1, Abstract 137, June (1983)

8   R. Langer, E. Edelman and D. Hsieh, Magnetically Controlled Polymeric Delivery Systems, Chapter 25, pp 585-596 in "Biocompatible Polymers, Metals, and Composites", M. Szycher ed., Technomic Pub. Co. (1983)

9   D. Hsieh, R. Langer and J. Folkman, Magnetic Modulation of Release of Macromolecules From Polymers, Proc. Natl. Acad. Sci. USA, 78:1863-1867 (1981)

10 J. Heller and R.W. Baker, Theory and Practice of Controlled Drug Delivery from Bioerodible Polymers, pp 1-17, in "Controlled Release of Bioactive Materials", R.W. Baker ed., Academic Press, NYC (1980)

11 C.G. Pitt, M.M. Gratzl, G.L. Kummel, J. Surles and A. Schindler, Aliphatic Polyesters II. The degradation of poly(DL-lactide), poly(ε-caprolactone), and their copolymers in vivo, Biomaterials, 2:215-220 (1981)

12 C.C. Chu, Hydrolytic Degradation of Polyglycolic Acid: Tensile Strength and Crystallinity Study, J. Appl. Poly. Sci., 26(5),1727-24 (1981)

13 H.B. Hopfenberg, Controlled Release from Erodible Slabs, Cylinders and Spheres, in "Controlled Release Polymeric Formulations", D.R. Paul and F.W. Harris eds., ACS Symposium Series No. 33, pp 26-32 (1976)

14 J. Heller, D.W.H. Penhale, R.F. Helwing, and B.K. Fritzinger, Release of Norethindrone from Polyacetals and Poly(OrthoEsters), Polym. Eng. Sci., 21:727-731 (1981)

15 A. Conix, Aromatic Polyanhydrides A New Class of High Melting Fiber-Forming Polymers, J. Polym. Sci., 29:343-353 (1958)

16 A. Conix, Poly[1,3-Bis(p-caorboxy-phenoxy)propane anhydride], Macromolecular Synthesis, pp 95-99, Vol. 2, J.R. Elliot ed., Wiley, NYC (1966)

17 H.B. Rosen, J. Chang, G.E. Wnek, R.J. Linhardt and R. Langer, Bioerodible Polyanhydrides for Controlled Drug Delivery, Biomaterials, 4:131-133 (1983)

ORAL SUSTAINED RELEASE DRUG DELIVERY SYSTEM

USING POLYMER FILM COMPOSITES

Sumita B. Mitra

Central Research Process Technolgies Laboratory
3M
St. Paul, MN   55144

INTRODUCTION

Of the several possible routes of introduction of sustained
release medication into the body the oral administration of single
dose medicinals is one of the simplest and safest since it does not
pose a sterility problem and the risk of damage at the site of
administration is minimal. Ordinarily, however, an oral sustained
release (SR) formulation is subjected to frequently changing
environments during transit through the gastro-intestinal tract as
it passes from the strongly acidic to the weakly alkaline medium in
the lower part of the small intestine. The variable absorbing
surfaces over the length of the GI tract adds further constraint to
the design of oral dosage forms. Moreover, the stomach emptying
time varies from patient to patient. These factors combine to
introduce considerable variability in the performance of oral SR
systems. In order to overcome these disadvantages, the dosage form
of this work was designed so as to be retained in the stomach for
prolonged periods of time and to release most of the active agent
under the controlled conditions therein at a steady rate. Several
approaches have been taken in the past to prolong the retention of
the dosage form in the stomach (1,2,3). This paper describes a
flexible, sustained release polymer-drug delivery composite film
device for oral administration which could be dispensed and
administered in a compact form which extends in the stomach to
remain buoyant. The film was suitably marked for facile measurement
of prescribed medical dosage according to length.

293

DESIGN OF SYSTEM

The drug delivery system described herein was a multilayered polymeric composite film consisting of two essential components. The first of these was a carrier film containing the active agent dispersed or dissolved in the matrix. The drug could be dispersed homogeneously in the matrix, but a better control of rate was obtained when the concentration of the medicament was increased from the outer wall to the interior of the film (Fig. 1). The second, essential component was the barrier film(s) which was placed on one or both surfaces of the carrier film. Ordinarily it did not contain any drug but served to control the rate of release of the active agent. In addition, the barrier film provided buoyancy of the drug delivery system by entrapping air in small pockets or "bubbles" between it and the carrier film (Fig. 2).

The choice of polymers suitable for forming the carrier film matrix and the barrier film were dictated by several factors among which were:

    (i)   compatibility with the gastric environment,
   (ii)   polymer stabilty during time of complete drug delivery,
 (iii)   appropriate mechanical properties i.e., capability of forming self-supporting films when containing the drug at high loading,
  (iv)   no appreciable swelling in water and having a softening point above 37°C,
   (v)   ease of fabrication
  (vi)   cost.

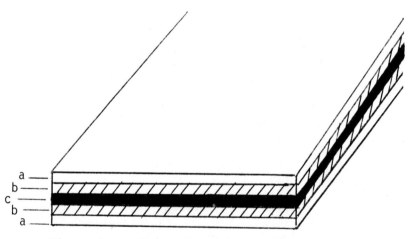

Figure 1. Multilayer carrier film
             (a)   16% quinidine glutonate (QG)
             (b)   37% QG
             (c)   51% QG

Figure 2. Top view of polymer-drug delivery film. The white areas indicate sealed air pockets.

Several polymers were found to fit all or most of the above criteria and were used to prepare the carrier films. Examples of polymers that were used as ethyl cellulose, poly(γ-benzyl-glutamate), poly(vinylacetate), cellulose acetate phthalate, and copolymer of methylvinylether with maleic anhydride. Table 1 summarizes some of the mechanical properties and sources for these polymers. Among the cellulose ether derivatives ethylcellulose (Ethocel -70 or 45, Dow Chem Corp) was found to be the material of choice.

In addition to the base polymers, plasticizers were often needed to impart a suitable degree of flexibility. Plasticizes which were found to be compatible with the polymeric materials of this investigation included acetylated monoglycerides, esters of phthalic acid such as dibutylphthalate and dioctylphthalate, D-sorbitol, diacetin, triacetin, dibutyltartarate, etc. An excipient was usually incorporated into the matrix of the carrier films. The excipients used were water-soluble materials which, as they dissolved, created channels in the polymer matrix and facilitated diffusion of the drug. Polyethylene glycol (PEG) of different molecular weights were used for this purpose.

The barrier or rate controlling films were in general somewhat more permeable to water than the carrier films. The materials used for this purpose consisted of a base, film-forming, water-soluble polymer in combination with at least one hydrophilic component such as hydroxypropylmethyl cellulose or polyvinylpyrrolidone. The polymers used were the same as those for the carrier film.

The composites of carrier and barrier films were generally cut into long narrow strips 2.1 x 14 cm^2. These were then packed into hard gelatin capsules. Several approaches were taken to ensure that the packed polymer-drug composite film would unfold once the encasing gelatin capsule dissolved. One method was to increase the hydrophilicity of one of the surfaces. The difference in degree of hydration caused the roll of film to open up in an aqueous medium.

Hydrophilicity was increased by incorporating a more hydrophilic polymer or by plasma deposition of a hydrophilic material on one surface. Alternatively, the film could be pleated in a corrugated form before inserting it into the capsule.

The drugs used in this work were the antiasthmatic agents theophylline and quinidine gluconate.

TABLE 1. Properties of Polymers and Polymer Films

POLYMER OR COPOLYMER	SOURCE	MOLECULAR WEIGHT[a] OR VISCOSITY[b]	FILM TENSILE STRENGTH PASCAL (P)
Ethyle cellulose	Dow	0.045-0.11 Pa.S, 0.063-0.085 Pa.S, (preferred for carrier film)	5.85×10^7
		0.041-0.085 Pa.S, (preferred for barrier film)	
Poly (γ-benzyl glutamate)	c	50,000-100,000[d]	1.65×10^7
Polyvinylacetate	3M[e]	50,000-100,000	3.8×10^7
Cellulose acetate phthalate	Eastman	50,000-100,000	f
Copolymer or methyl vinly ether - maleic anhydride	GAF	High viscosity type, Gantrez An-169	

(a) Number average molecular weight
(b) Viscosity of Methocel E-15 determined as a 2% solution in water at 20°C. Viscosity of Ethocel - 45 determined as a 5% solution in toluene-ethanol (80:20, vol/vol) at 25°C. Viscosity of Ethocel - 70 determined as a 5% solution in toluene-ethanol (60:40 vol/vol) at 25°C.
(c) Prepared according to directions of S. B. Mitra, N. K. Patel and J. M. Anderson, Int. J. Biol. Macromolecules 1, 55 (1979)
(d) Molecular weight determined in dichloroacetic acid[4]
(e) Scotchpak® heat sealable polyester film No. 112, 113, 115, 125
(f) Manufacturer's bulletin reports CAS Reg. 9004-38-0

EXPERIMENTAL

Dispersions of drugs were made up by adding them to solutions
of the carrier polymer in appropriate solvents such as acetone, 3:1
chloroform-methanol or 3:1 CH_2Cl_2-EtOH. Other components, when
needed, included a plasticizer such as Myvacet 9-40 (Eastman) and an
excipient such as PEG. The viscosity of the resultant mixture was
0.1-0.5 Pa.S. After homogenization, the dispersion was knife coated
to give dried films 4-10 mils in thickness. Multilayer films were
prepared by sequential coating of solutions of different
concentrations of drug. The barrier films (1 mil) were also
prepared by knife coating. Alternatively, the mixture of polymer,
drug, and plasticizer could be extruded into thin films. The
"bubble" films were prepared by means of an embossed die-containing
heat sealing metal apparatus. Sealing was performed at 135°C and
9.65×10^5 Pa for 0.75 seconds. The prepared films were cut to
required size (2.1×14 cm^2), pleated or rolled, and inserted into
size 0 hard gelatin capsules.

In vitro measurement of release rate profile was performed
using USP-2 dissolution system with 900 ml 0.1 N HCl. A Teflon[R]
screen was placed 2.5 cm from the bottom of the flask to prevent
contact of the paddle with the floating film. The dissolution
medium was monitored by UV at 270.5 nm for release of theophylline
and at 334 nm for quinidine gluconate. The apparent density of the
films were measured in water using a glass pycnometer. Gastric
residence times were measured in vivo using beagle dogs. The
barrier films were coated or streaked with radio-opaque materials
such as barium sulfate or tantalum dust and then administered with
50 ml of water to the dogs. Non-disintegrating radio-opaque tablets
served as controls. The dogs were x-ray examined every hour to see
the movement of film along the GI tract. Food and water were
available ad lib. For pharmaokinetic studies blood level of drug
was determined in fasted dogs.

Scanning electron micrographs were taken using an ISI Super-2
scanning electron microscope. The barrier films were initially
formulated with PEG of different molecular weights. Micrographs of
the films prior to and after soaking in the stirred dissolution
medium were recorded.

RESULTS AND DISCUSSION

In vitro studies on the composite films showed that while the
completely laminated films were not very effective in floating in
aqueous environment, the "bubble" type films were extremely buoyant.
A limiting condition for the films to be buoyant was determined to
be a value of 0.66 for the ratio of laminated or sealed to unsealed
area. Under these conditions, the overall apparent density of the

composite film structures was less than that of gastric juice (i.e., less than 1.0). Theophylline containing composite film structures of this work having the configuration shown in Fig. 2 were prepared having a variety of lengths and widths. Theophylline-containing composite films, which were similarly laminated but had no air pockets, served as controls. They were also prepared having the same variety of lengths and widths as shown in Table 2. The surface area and weight per unit area for all the composite structures were essentially the same, yet the apparent density of the controls was greater than that of the sustained release device of this work, wherein the apparent density is less than that of gastric juice. Furthermore, the control and the device of this work had the same type of wetting characteristics as determined by contact angle measurements; hence, wetting was not a factor in floatation behavior.

TABLE 2. Buoyancy Studies of a Theophylline-Containing Sustained Release Device[a]

Laminated Composite	Composite Measurements		
	Length x Width	Weight/Unit Area	Apparent Density[c]
Film Type[b]	(cm^2)	(g/cm^2)	(g/cm^3)
C	2.1 x 14	0.0233	1.21
SRD	2.1 x 14	0.0241	0.62
C	3.1 x 9.1	0.0235	1.18
SRD	3.1 x 9.1	0.0232	0.62
C	4.2 x 7.0	0.0234	1.24
SRD	4.2 x 7.0	0.0231	0.64
C	5.1 x 5.1	0.0230	1.24
SRD	5.1 x 5.1	0.0234	0.62
C	6.1 x 4.7	0.0220	1.23
SRD	6.1 x 4.7	0.0226	0.66

(a) In all cases, the thickness of the carrier film is 6 mil (0.015 cm) and that of the barrier films is 1 mil each (0.002 cm)
(b) C = Control; SRD = Sustained Release Device
(c) Determined with a pycnometer; density of gastric juices = 1.004-1.01

TABLE 3. Dependence of Pore Size on Molecular Weight of Excipient

MOL. WT. OF PEG	AVERAGE PORE SIZE (μ)
380- 400	0.5
6,000- 7,500	1.5
15,000-20,000	5.0

Scanning electron micrographs of barrier films after drug release was complete (Fig. 3) indicated that the pore sizes created in the barrier films were dependent on the molecular weight of PEG (Table 3). The dissolution of the excipient evidently created porous regions into which water penetrated and triggered the release of the active agent. This mechanism occurred concurrently with the diffusion of the drug through the matrix.

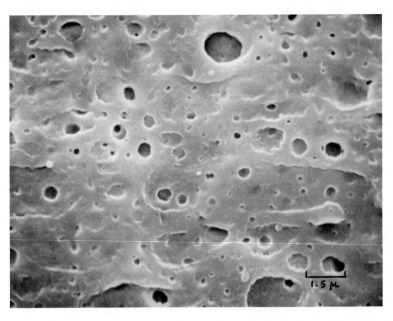

Figure 3. SEM of carrier film after dissolution of excipient

The release behavior of drug from the films were found to be dependent on the following physical parameters:

i)   thickness of barrier film
ii) composition of barrier film
iii) initial drug concentration of carrier films
iv) gradient of concentration of drug in carrier film
v)   molecular weight of barrier film polymer
vi) hydrophilicity/hydrophobicity ratio of barrier film
vii) pH of external media

As an example, Figure 4 (A-E) shows the dependence of release profile on the molecular weight and composition of the barrier film. In each case, the composition of the carrier film was maintained constant (51% theophylline). The results are summarized in Table 4. The results indicate that, while higher proportions of the low molecular weight ethocel 45 were effective in decreasing the period of release, considerable departure from zero-order behavior occurred at the same time. We have found, however, that by increasing the hydrohilicity of the barrier film compositions containing greater proportions of ethocel-70 it was possible to approach zero-order behavior while at the same time maintain a short release period. This is illustrated in Figure 5A for an outer film made from ethocel-70/ethocel/45/methocel E-15. (Table 5 gives compositions of this film.) The value of R^2 increased to 0.921 in this case.

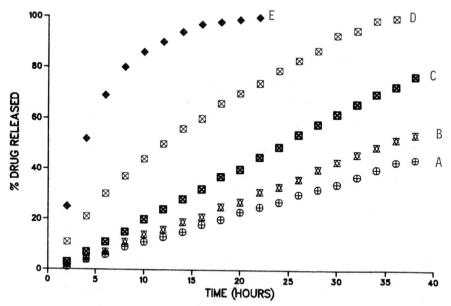

Figure 4. Effect of viscosity of Ethocel of carrier on release rates

TABLE 4.   Dependence of Release Behavior of Theophylline on
           Viscosity of Ethocel (Ref. Figure 4)

Example	Ethocel 70:Ethocel 45 (w/w)	1/2(b) hrs.	R^2(a)
A	100:0	42	0.999
B	90:10	35	0.999
C	70:30	25	0.999
D	50:50	10	0.976
E	30:70	3	0.876

(a)   For zero-order reaction.
(b)   Half life

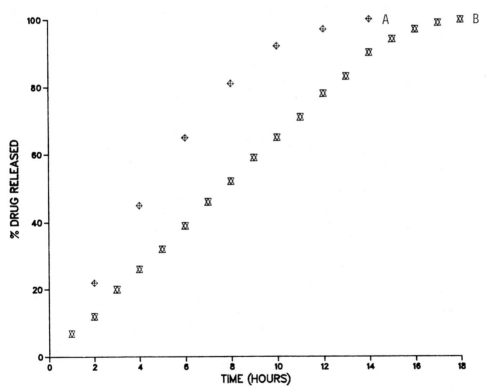

Figure 5.   (A)   Release of Theophylline from barrier film on
                  Ethocel-70/Ethocel-45/Methocel E-15.
            (B)   Release of Quinidine Gluconate from multilayer
                  carrier film

301

TABLE 5. Composition of Carrier and Barrier Films of Fig. 5A.

Materials	Carrier Film (Wt %)	Barrier Film (Wt %)
Ethocel-45	--	21
Ethocel-70	24.6	21
PEG-380[a]	12.4	22
Myvacet-9-40	11.6	19
Methocel-E-15[b]	--	17
Theophylline	51.4	--

(a) polyethylene glycol, average molecular weight 380-400
(b) Dow Chemical Corporation, Midland, MI

The use of multilayer carrier film construction (Table 6), where the concentration of the drug increased towards the interior of the film, resulted in zero-order rate profile over almost the entire period of release. This is illustrated in Fig. 5B which shows the release of quinidine gluconate from multilayer carier film of Fig. 1. ($T_{1/2}$ = 7.5 hrs., R^2 = 0.988).

TABLE 6. Composition of Carrier Film of Figure 5B

Materials	Layers		
	(a) wt %	(b) wt %	(c) wt %
Ethocel-70	40.1	28.3	24.3
Myvacet 9-40	18.9	13.3	11.6
PEG-380	20.2	14.2	12.4
Quinidine Gluconate	20.8	44.2	51.7

In vivo gastric residence time determinations were performed on 19 beagle dogs. The results of individual experiments are summarized in Table 7. The results indicated that the sustained release device did open up or unfold (as seen by x-ray examination) in the stomach of beagles and that gastric residence time (mean = 6.5 hrs.) was considerably longer than that of the controls (mean = 2.5 hrs.). Pharmacokinetic studies showed that the bioavailability of these drugs in vivo was constant during the release period.

TABLE 7. In Vivo Gastric Residence Time Determination of
         Envelope Films

	Gastric Residence (hrs)	
Expt. No.	Film	Control Tablet
1	7	—
2	5	1
3	4	1
4	8+[a]	6
5	7	4
6	8+[a]	1
7	8	2
8	6	1
9	5	2
10	4	4
11	6	1
12	5	3
13	8+[a]	2
14	7	4
15	8	—
16	6	2.5
17	—	1
18	8+[a]	2
19	7	5
x̄ =	6.5	2.5
s.d. =	1.38	1.61

(a) In these cases the film was present in the stomach after
     8 hours. Expt. was discontinued at this time.

REFERENCES

1. A. S. Michaels, A. Zaffaroni, U.S. Pat. 3,901,232
    R. E. Mamajek, E.S. Moyer, U.S. Pat. 4,207,890
    R. H. Johnson, E. L. Rowe, U.S. Pat. 2,574,820
    G. S. Banker, Brit. Pat. 1,428,426
2. S. Watanabe et. al., U.S. Pat. 3,976,764
3. P. R. Sheth and J. L. Tossounian, U.S. Pat. 4,140,755; 4,167,558
4. P. Doty, J. H. Bradbury, and A. M. Holtzer, J. Amer. Chem. Soc.
    78, 947 (1956).

CHEMICAL CHARACTERIZATION OF AN IMMOBILIZED HEPARIN: HEPARIN-PVA

Cynthia H. Cholakis and Michael V. Sefton

Department of Chemical Engineering & Applied Chemistry

University of Toronto, Toronto, Ontario, M5S 1A4

INTRODUCTION

Heparinization of polymers[1,2], either ionically or covalently, has been one of the techniques used with varying degrees of success to impart thromboresistance to materials. For example, the covalent coupling of heparin to polyvinyl alcohol (PVA) has resulted in a heparin-PVA hydrogel which appears to retain blood compatibility in vitro[3,4], in the absence of significant heparin release. It is the aim of this paper to report on the chemical characterization of the heparin-PVA hydrogel in terms of heparin content and point of coupling. The complexity and heterogenity of heparin make this a difficult task for this or any heparinized biomaterial.

Commercial heparin preparations isolated from either lung or intestinal mucosal tissue are composed of single polysaccharide chains with molecular weights ranging from below 10,000 to well over 20,000. More than 70% of the structure of "conventional" heparin can be accounted for by repeating disaccharide units consisting of 1,4-linked L-iduronic acid and D-glucosamine residues (Figure 1). The iduronic acid residues are O-sulfated at position 2, and the glucosamine residues are N-sulfated and O-sulfated at position 6.

Heparin chains are present in tissues as proteoglycans, i.e., bound to a protein from which they are released, during their isolation, by the combined action of proteolytic enzymes and base. Consequently, commercial heparin preparations still, depending upon their mode of production, contain a linkage region consisting of β-linked glucuronic acid, two galactose residues, an xylose residue and serine (Figure 2). Heparin which has been prepared by

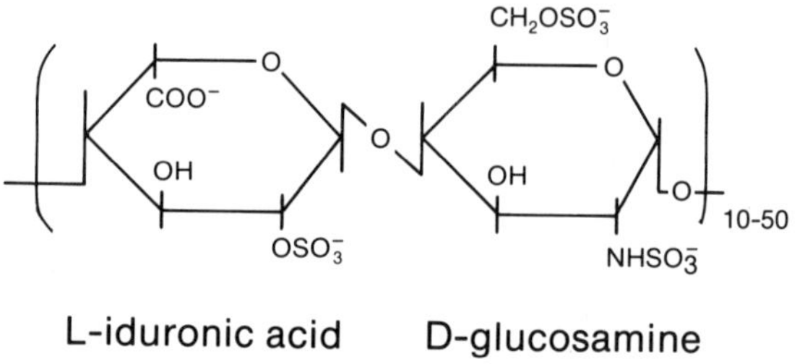
CH2OSO3-
O
COO-
O
OH
OH
O
10-50
OSO3-
NHSO3-
L-iduronic acid
D-glucosamine

Fig. 1. Typical repeat structure of heparin.

mild methods, essentially based on treatment with proteolytic enzymes, contain residual amino acids with serine as the main component. The absence of the linkage region in some commercial heparin preparations has been attributed to oxidizing agents utilized late in the isolation procedure to "bleach" the heparin.

Amino acid analysis of various commercial preparations of heparin has demonstrated a large variation in the amino acid content. Wang and Jaques[5] reported an amino acid content in commercial heparins between 0.3 to 2.3% (w/w) with no correlation with anticoagulant activity. The lack of consistency in the amino acid compositions of these preparations suggests that the presence of particular amino acids is due to differences in conditions in the various extraction and purification processes. It is suggested that the impurities may copurify as counter ions to the highly acidic heparin chain.

Preparation of the heparin-PVA hydrogel involves a cross-linking reaction of the hydroxyls of polyvinyl alcohol with glutaraldehyde producing an acetal bridge (Figure 3). Acetalation

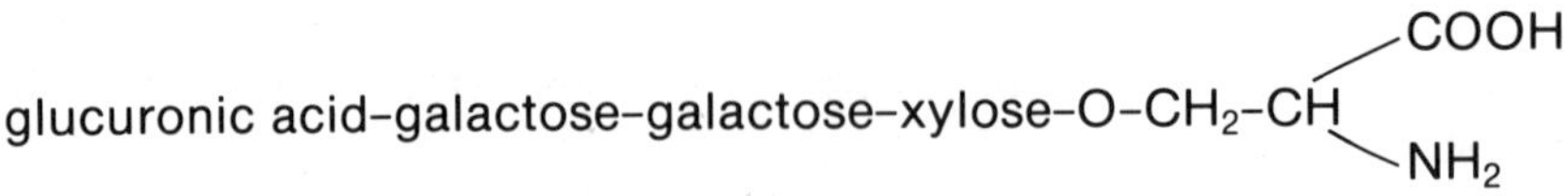
COOH
glucuronic acid-galactose-galactose-xylose-O-CH2-CH
NH2

Fig. 2. Linkage region of heparin to remainder of proteoglycan (14).

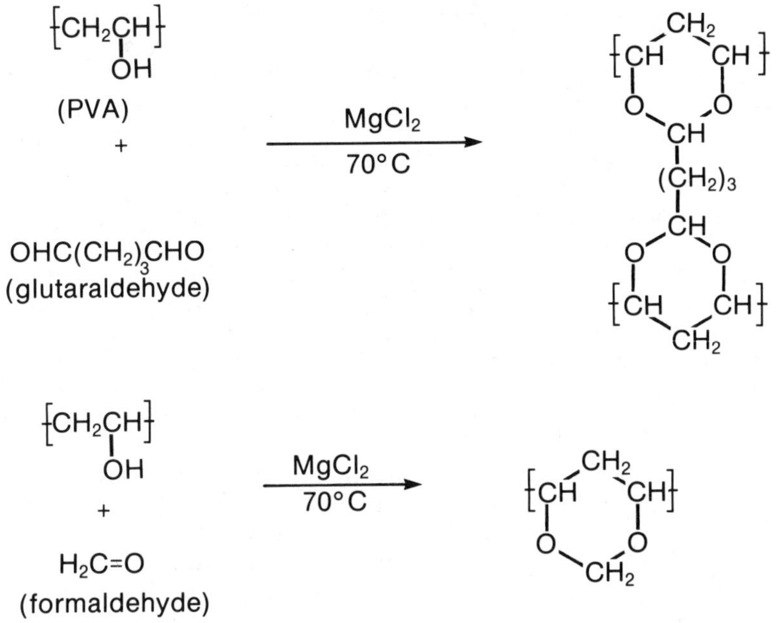

Fig. 3. Acetal bridge and cyclic acetal products of reaction of polyvinyl alcohol with glutaraldehyde and formaldehyde.

of adjacent pairs of hydroxyls without crosslinking occurs with formaldehyde. Although similar acetal rings could be involved in heparin immobilization (to the secondary hydroxyls of heparin) as was proposed earlier[3], the terminal serine or other free NH_2 group was thought to be a more likely site of binding. One objective of this work was to test this hypothesis.

MATERIALS AND METHODS

Heparin-PVA films: Standard heparin-PVA solutions consisted of 10% (w/w) polyvinyl alcohol (20% acetylated, Gelvatol 20-60, Monsanto Canada Ltd., Toronto, Canada), 5% Magnesium chloride (Lewis acid catalyst), 2% glutaraldehyde, 3% formaldehyde, 4% glycerol, 1% sodium heparin and water. The heparin-PVA solutions

307

were cast onto glass petri dishes, dried overnight at room temperature and cured at 70°C for two hours to effect cross-linking. The heparin-PVA gels were reswollen and stored in 0.05 M sodium phosphate buffered 0.15 M NaCl (PBS, pH 7.4).

Heparin: Four commercial heparins were used: porcine intestinal mucosal heparin (173 USP units/mg, Canada Packers Ltd., Toronto); a bovine/porcine mixture (120 USP units/mg, supplied by Dr. D. Ives, Connaught Laboratories Ltd., Toronto); porcine heparin, 195 USP units/mg, Diosynth, Oss-Holland); and porcine intestinal mucosal heparin (164 USP units/mg, Sigma, Saint Louis, MO).

Modified heparin: Tritiated heparin was obtained from Dr. M.W.C. Hatton, of McMaster Unversity, Hamilton. The labelling procedure[6] incorporated tritium onto the heparin molecules possessing a terminal monosaccharide (i.e., those that did not terminate in serine) which on reduction with NaB^3H_4 were reduced to yield an alditol.

Complexes of Connaught heparin with amino acids were prepared by contacting heparin with Dowex 50W-X(H^+) cation exchange resin (J.T. Baker Chem. Co., Phillipsburg, NJ) to remove sodium and then collecting the heparin in an aqueous amino acid solution containing no more than 3 mole amino acid/mole heparin. Glycine (BIORAD Laboratories, Richmond, CA), L-(+)-aspartic acid, L-(+)-glutamic Acid (Fisher Scientific Co., Fair Lawn, NJ) and L-(+)-lysine monohydrochloride (Eastman Kodak Co., Rochester, NY) were individually complexed with heparin.

Heparin characterization: The N-sulfate content of heparin was determined by measurement of the inorganic sulfate content liberated by deamination with nitrous acid. The inorganic sulfate was determined turbidimetrically as barium sulfate[7].

Heparin (approximately 2 mg/sample) after hydrolysis in 6 N HCl under vacuum was analyzed for amino acid content (Faculty of Medicine, Amino Acid Analyser, University of Toronto).

Heparin content and release rate: The stability and heparin content of heparin immobilized on to PVA was determined using the toluidine blue/metachromatic shift assay[4]. The heparin-PVA films were washed for up to 200 hours with PBS. The wash solution was analyzed for heparin by adding toluidine blue and measuring the change in absorbance at 600 nm (Beckman Model 25 UV/VIS Spectrophotometer, Fullerton, CA). The heparin content of the wash solution was determined by comparing the change in absorbance to a standard curve, prepared with known amounts of the same heparin. The detection limit of this assay was ~ 2 ppm (w/w) heparin. The release rate was determined from the slope of a plot of heparin concentration against time as before[4], knowing the volume of the wash solution and the weight of the gel. The heparin content of the heparin-PVA gel was then determined indirectly by mass balance.

The wash solutions were also analyzed spectrophotometrically for heparin via a glucosamine analysis using indole[8] and a uronic acid analysis using carbazole[9].

The incorporation of tritiated heparin was monitored by adding 10 mL Aquasol (New England Nuclear) to 2 mL samples of wash solution or 0.3 g samples of gel and counting for 10 minutes on a Beckman LS 8000 liquid scintillation spectrometer.

RESULTS

The amount of heparin bound to the gel was correlated approximately with the amount of terminal serine in the four commercial heparin samples examined (Table 1): the higher the serine content the greater was the amount bound to the gel after 1000 hours of washing in PBS.

The tritiated heparin added to the original heparin-PVA solution was not incorporated into the gel. Regardless of the glutaraldehyde concentration, $99.48 \pm 0.09\%$ of the total counts were found in the wash solution within a few days. Thus almost all of the chains terminating in a reducing sugar (i.e., the labelled molecules) did not bind to the gel. Approximately 50% of the unlabelled heparin was found to be bound under the same conditions by toluidine blue assay.

There is, as shown in Table 2, a substantial difference in amino acid content, especially serine between the heparin incorporated into the PVA-gel solution and that recovered by lyophilization of the wash solution. Commercial heparin from Canada Packers Ltd. and Sigma Chemical Co. had serine contents of 30.24 and 14.92 nmole/mg heparin compared to 2.92 and 5.27 nmole/mg for the unretained heparin, respectively.

Table 1. Effect of heparin source on final heparin content.

Heparin Source	Heparin Content (% original)[*]	Serine Content of Heparin[‡] (n mole/mg)
Canada Packers	62 (± 2)	38
Connaught	32 (± 14)	8
Sigma	26 (± 2)	6
Diosynth	25 (± 3)	1

[*] fraction (x100) of heparin added to gel solution that was immobilized (i.e., not found in the wash solution) as determined by mass balance.

[‡] not corrected for the 5-15% loss of serine known to occur with acid hydrolysis (15).

Table 2.   Amino acid analysis of heparin and heparin recovered
          from the wash solution by precipitation.

Amino Acid	Canada Packers		Sigma	
	Heparin	Ppt. Heparin	Heparin	Ppt. Heparin
Aspartic acid	2.43	0.81	1.66	1.15
Threonine	1.21	–	–	–
Serine	30.24	2.92	14.92	5.27
Glutamic acid	1.98	1.74	2.66	1.65
Glycine	1.86	2.58	2.10	3.79
Alanine	–	1.00	–	–

units are n mole/mg

Connaught heparin complexed with various amino acids were
significantly more reactive towards PVA than commercial heparin
(Table 3). The heparin-lysine complex was the most reactive, with
96.6% of the original heparin binding after 30 days compared to
13% for unmodified heparin. The acidic amino acid heparin
complexes were second most reactive and glycine-heparin the least
reactive. The elution rate of heparin into the wash solution
varied for each complex and degree of complexation but was the
same order of magnitude as that for uncomplexed heparin (1.67 x
10^{-2} µg/gwet gel•min).

Varying the heparin content in the original PVA solution from
0.5% to 2% did not increase the percentage of heparin immobilized.
With 2% glutaraldehyde in the original PVA solution, 79.6 ± 1.2%
of the initial heparin was bound to the gel (after 150 hours of
washing) according to the toluidine blue assay. However,
measurement of the wash solution by glucosamine and uronic acid
assays revealed a discrepancy in the heparin concentration (Table
4). The heparin content in the wash solution (i.e., unretained
heparin) appears to have been underestimated by the toluidine blue

Table 3.   Binding of amino acid heparin complexes to PVA.

Amino Acid	Heparin Content % original (after 1 month)
Glutamic acid	58.8
Aspartic acid	48.0
Glycine	28.9
Lysine	96.6
Heparin (control)[*]	13.0

*Connaught heparin, 0.5% glutaraldehyde, 1% heparin.

Table 4.   Unretained heparin concentration in a wash solution
determined by various assays.

Assay	Heparin Concentration (ppm)
toluidine blue	25
indole (glucosamine)	47
carbazole (uronic acid)	43

assay by a factor of almost two. The unretained heparin was found
also to have a 30% lower N-sulfate content than the starting
material, presumably as the result of the curing process.

DISCUSSION

Based on the correlation with serine content, the lower
serine content in the unretained heparin and the absence of
binding for the tritiated heparin molecules, which by virtue of
the tritiation process, did not contain serine, a description of
the primary point of coupling of heparin to PVA has been
postulated (Figure 4).   The terminal serine resulting from
cleavage of the proteoglycan during commercial production appears
necessary for binding of heparin by PVA. This is an important
result since heparin bound in this way is presumably free to
maintain its native conformation leaving its active sites intact
and free to interact with thrombin and antithrombin III.   This
accounts for the observed thromboresistance[3,4].

The serine content of Canada Packers heparin varied from lot-
to-lot but, assuming a molecular weight of 13,000, a serine
content of 38 nmole/mg heparin, translates to approximately 50% of
the chains terminating in a serine residue. Thus binding through
this source can account for about 50% immobilization of the
original heparin. However, apparent immobilization of up to 80%
of the heparin has been observed.   This discrepancy can be
resolved in large part by the error associated with the toluidine
blue assay.   Since the immobilized heparin contents were
determined by mass balance from the solution values and assuming a
factor of two underestimation of concentration over the entire
concentration range by the toluidine blue assay (Table 4), the
heparin content in the gel may more accurately be 60% for gel
prepared with 2% glutaraldehyde (i.e., 40% of the heparin in the
wash solution rather than the 20% found by toluidine blue).

The mechanism of reaction between the amino group of serine
and glutaraldehyde is not clear. The expected interaction leads
to a simple Schiff base although other reaction mechanisms for
glutaraldehyde have been proposed. If the reactive species are α-
β-unsaturated aldehydes formed by aldol condensation of

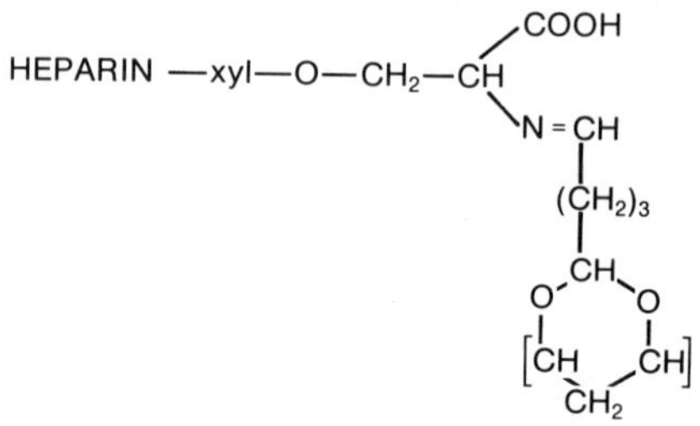

HEPARIN —xyl—O—CH$_2$—CH
$\diagup$COOH
$\diagdown$N = CH
$|$
(CH$_2$)$_3$
$\diagup$CH$\diagdown$
O$\qquad$O
$[$CH$\qquad$CH$]$
$\diagdown$CH$_2$$\diagup$

Schiff base

Fig. 4. Possible structure of bond linking terminal serine of
heparin to polyvinyl alcohol. This structure may be
subject to further rearrangement.

glutaraldehyde during the curing process, Michael type additions
or resonance stabilized Schiff base formation can occur. The
Schiff base shown in Figure 3 may also rearrange to a more stable
structure. Spectroscopic investigations are underway to better
define the binding site structure.

The 10% of the binding that remains to be accounted for, has
been attributed to binding through the contaminant amino acids,
since 10% of the Canada Packers heparin chains contain, on
average, an amino acid other than serine[4]. All amino-acid-heparin
complexes were more reactive towards the gel but the lysine-
heparin complex was the most reactive presumably because it
possesses two free amino groups. One of the amino groups of
lysine can interact with the negatively charged heparin chain
while the other amino group can bind to PVA via glutaraldehyde.
Steric effects are presumed to prevent substantial binding to the
few free glucosamine residues in commercial heparins; this has
been confirmed experimentally[10].

Assuming a similar factor of two error in the toluidine blue
assay, the heparin release rate should be twice that reported in
the results section or earlier[4]. Nevertheless, this release rate
is still substantially less than the minimum reported necessary
for thromboresistance of ionically linked heparin[11,4] or that
necessary to create a heparin microenvironment at the blood
material interface[12].

312

The error in the toluidine blue assay, reflects the difference between the heparin released into the wash solution and that used to prepare the orginal calibration curve (i.e., the starting heparin). This difference was evident in the lower N-sulphate content in the unretained heparin relative to the starting heparin. Since the degree of metachromatic shift associated with the toluidine blue-heparin complex[13] is directly related to the N-sulphate content of the heparin, the toluidine blue assay error was attributed in part to the lower N-sulphate content of the unretained heparin and the corresponding use of too high a slope of the calibration curve (plot of change in absorbance against heparin concentration) for the unretained heparin.

Another reflection of heparin's heterogeneous nature, is the previously reported[4] difference between toluidine blue determined heparin contents and those determined by sulphur-35 labelled heparin. ^{35}S-heparin severely underestimated the heparin content (1.1 mg/g gel film) relative to that determined by toluidine blue (14.4 mg/gel). Although this was attributed to loss of label during preparation of the gel, this descrepancy is now presumed to reflect a lower serine content in the commercially obtained (Amersham) radiolabelled heparin which would not be representative of the unlabelled heparin. Thus extrapolation of measurements based on one particular heparin, here ^{35}S-heparin, to a second heparin (e.g., Canada Packers heparin) is hazardous at best.

Although heparin is released from the gel too slowly to account for the observed thromboresistance[4], the observation that the release rates of amino acid-heparin complexes are similar to that of the unmodified heparin complicates efforts at determining the useful life of the material. It is interesting to speculate that the observed release with unmodified heparin is due to release of heparin bound ionically via the contaminant amino acids and not due to release of heparin bound via the terminal serine. If this were true then the release rate should eventually decrease to that characteristic of the covalently linked heparin (which may well be zero). Thus the useful life of the material could well exceed that extrapolated from the release rates observed after 1-2 weeks. A progressive decrease in the release rate has been observed between 300 and 1000 hours of washing[4] but this may still not be a long enough period for all of the ionically linked heparin to have been removed.

CONCLUSION

The absence of binding for the tritium labelled heparin, without serine, demonstrated that the terminal serine is the predominant site of binding for heparin immobilized to PVA with glutaraldehyde. This heparin, though covalently bound, retains its activity presumably because it is free to maintain its native conformation. However, quantifying the retention and elution rate of immobilized heparin is complicated by its heterogenous nature and particularly by differences between the immobilized and unretained heparin and between the unretained and starting heparin. The toluidine blue assay technique assumes that the heparin washed from the gel interacts with toluidine blue to the same extent as the original, heparin, which does not hold for heparin-PVA. Thus in this case determination of heparin content or release rate is erroneous and heparin concentration should be determined via its glucosamine or uronic acid content. Similarly the use of ^{35}S-heparin or ^{3}H-heparin is only valid if the labelled heparin's serine content is that of the unlabelled heparin. Careful attention to the potential effect of heparin's heterogenerty and the characterization of immobilized heparin is essential to interpret the results properly, regardless of the assay used.

ACKNOWLEDGEMENTS

The authors acknowledge the financial support of the National Institutes of Health (HL 24020) and the Ontario Heart Foundation. CHC is grateful to the Natural Sciences and Engineering Research Council for a postgraduate scholarship.

REFERENCES

1. S.W. Kim, C.D. Ebert, J.Y. Lin, J.C. McRea, Nonthrombogenic polymers, pharmaceutical approaches. ASAIO J. 6:76–87 (1983).
2. S.W. Kim, C.D. Ebert, J.C. McRea, The biological activity of antithrombotic agents immobilized on polymer surfaces, Ann. N.Y. Acad. Sci. (in press).
3. M.F.A. Goosen and M.V. Sefton, Heparinized styrene-butadiene-styrene elastomers, J. Biomed. Mater. Res. 13:347 (1979).
4. M.F.A. Goosen and M.V. Sefton, Properties of a heparin-poly(vinyl alcohol) hydrogel coating, J. Biomed. Mater. Res. 17:359–373 (1983).
5. C.C. Wang and L.B. Jaques, The amino acid content of heparin, Arzneim-Forsch (Drug Res.) 24,:Nr.12 (1974).
6. M.W.C. Hatton, L.R. Berry, R. Machovich, E. Regoeczi, Tritiation of commercial heparins by recation with NaB^{3}H$_{4}$: chemical analysis and biological properties of the product, Anal. Biochem. 106:417–426 (1980).
7. Y. Inoue and K. Nagasawa, A new method for the determination of N-sulfate in heparin and its analogs, Anal. Biochem. 71:46–52 (1976).

8. D. Lagunoff, G. Warren, Determination of 2-deoxy-2-sulfoaminohexose content of mucopolysaccharides, Arch. Biochem. Biophys. 99:396-400 (1962).
9. T. Bitter, H.M. Muir, A modified uronic acid carbazole reaction, Anal. Biochem. 4:330-334 (1962).
10. C.H. Cholakis, (unpublished results).
11. H. Tanzawa, Y. Mori, N. Harumiya, H. Miyama, M. Hori, N. Oshimu and Y. Idezuki, Preparation and evaluation of a new athrombogenic heparinized hydrophilic polymer for use in cardiovascular system, Trans. ASAIO 19:188-194 (1973).
12. D. Basmadjian and M.V. Sefton, Relationship between relese rates and surface concentration for heparinized materials, J. Biomed. Mater. Res. 17:509-518 (1983).
13. A. Wollin and L.B. Jaques, Metachromasia: an explanation of the colour change produced in dyes by heparin and other substances, Thromb. Res. 2:377-382 (1973).
14. B. Casu, Structure and biological activity of heparin and other glycosaminoglycans, Pharm. Res. Com. 11:1-18 (1979).
15. U. Lindahl, J.A. Cifonelli, B. Lindahl, L. Roden, The role of serine in the linkage of heparin to protein, J. Biol. Chem. 240:2817-2820 (1965).

TUMORCIDAL ACTIVATION AND KINETICS OF ECTOENZYME PRODUCTION

ELICITED BY SYNTHETIC POLYANIONS

Raphael M. Ottenbrite and Kristine Kuus

Department of Chemistry and Microbiology, and the
    Massey Cancer
Virginia Commonwealth University
Richmond, Virginia 23284

Alan M. Kaplan

Department of Microbiology and Immunology
School of Medicine
University of Kentucky
Lexington, Kentucky 40536

Polyanionic compounds have been shown to exert a profound influence on the immune response (1,2). These compounds are capable of modulating responses of T lymphocytes, B lymphocytes, and macro-phages when administered via the appropriate route and at the appropriate time. They have been shown to enhance and suppress specific and non-specific immune responsiveness (3). Pyran, a polyanionic compound composed of maleic anhydride and divinyl ether, exerts its effect on T cells, B cells, and macrophages (1). The fact that pyran can "activate" macrophages to nonspecifically kill tumor cells has been the subject of a great deal of investigation. The role of activated macrophages as a mechanism of immunopotentiator-mediated anti-tumor activity has been established. However, little information exists with respect to the factors which control macrophage differentiation to the activated state. Certainly, significant differences in the properties of normal, elicited and activated macrophage have been demonstrated. However, the signals that trigger the events which lead to these states are unknown.

Polyanionic compounds similar to pyran with differing molecular weight, hydrophobicity, and charge density were synthesized and their ability to non-specifically "activate" macrophages was evaluated. The level of activation of test polymer-elicited macrophages was monitored by their nonspecific tumoricidal capacity and ectoenzyme profiles. In addition, the ability of test polymers to inhibit growth of Lewis lung in vivo was evaluated as well as polymer-induced changes in the number and composition of peritoneal exudate cells and peripheral leukocytes.

We have determined that the activation of macrophages by test polymers differs from that of the conventional activating agents pyran and C. parvum. The kinetics of cytotoxicity ectoenzyme profiles, peritoneal exudate cells and peripheral leukocyte responses and in vivo tumoricidal activities of test polymer-elicited macrophages did not correlate with the activity patterns observed with pyran or C. parvum.

The major objectives of this study are to further define the chemical requirements for macrophage activation by synthetic polymers that vary in lipophilicity, surface charge, and molecular weight and to evaluate the activation state of macrophage elicited with test compounds in order to either eliminate or corroborate the contribution that each of these variables make in the acquisition of the "activated" state.

Introduction of Cytotoxic Macrophage by Polyanionic Compounds

A series of 10 polyanionic polymers (Table 1) as well as pyran, C. parvum, thioglycollate and nothing were tested for their ability to elicit peritoneal macrophages with tumoricidal capacity. Peritoneal exudate cells (PEC) from mice that received 50 mg/kg test polymers of pyran, or 17.25 mg/kg of C. parvum i.p. were collected on day 7, PEC from mice receiving 1 mL of a 10% solution of Brewer's thioglycollate i.p. were collected on day 3. PEC pooled from five mice per group were then tested in morphologic and ^{3}H release assays against Lewis lung. Results of these experiments are shown in Table I.

A good correlation was obtained for the two assays of the test polymers. IAS, CDA-MA and MP-MA demonstrate the greatest capacity for induction of tumoricidal macrophage in both assays. Each of these polymers possess a hydrophobic group and enhanced surface charge properties. LS-MA which is a very low molecular weight (MW ~ 1600) of styrene-maleic anhydride copolymer contains a benzene ring, did not induce cytotoxic macrophages. MAVA and EA did not induce PEC to cytotoxic potential though their surface charge properties are enhanced.

Peritoneal exudate cells (PEC) induced by test polymers as well as controls demonstrate a dose response in their cytotoxicity to Lewis lung. Essentially no cytotoxicity was seen at E:T ratios of 1:1 and the most significant differences among the groups were discerned at an E:T ratio of 20:1. Though the two cytotoxicity assays differ somewhat in their sensitivity, it may

be concluded that CDA-MA, MP-MA and IAS are the most biologically active polymers. Thus, it would appear that CDA-MA, MP-MA
and IAS induce PEC nonspecific tumoricidal activity in a manner
analogous to pyran and C. parvum. Test polymer, as well as
pyran and C. parvum-induced PEC are not cytotoxic to newborn
mouse fibroblasts (data not shown).

Ectoenzyme Profiles of Polymer Elicited Macrophages

Macrophage ectoenzyme profiles, consisting of 5'-nucleo-
tidase (5'N) Leucine Aminopeptidase (LAP) and Alkaline Phospho-
diesterase (APD), have been reported to discriminate between
resident and thioglycollate, pyran or C. parvum-induced macro-

Table 1. Introduction of Cytotoxic Macrophages with polyanionic
polymers.

Polyanionic Polymer	Symbols	% Cytotoxicity Morphological Assay
Control	--	10
Divinyl Ether Maleic Anhydride	Pyran	80
C. Parvum	--	82
Poly(itaconic acid-co-stryene)	IAS	84
Poly(vinyl acetate-co-maleic anhydride)	VAMA	50
Poly(cyclohexyl-1,3-dioxepin-co-maleic anhydride)	CDA-MA	80
Poly(allylurea-co-maleic an-hydride)	AU-MA	5
Poly(4-methyl-2-pentenone-co-maleic anhydride)	MP-MA	75
Poly(ethylene-co-maleic an-hydride)	E-MA	12
Poly(ethacrylic acid)	EA	40

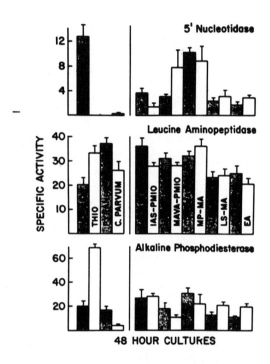

Figure 1   Ectoenzyme Activity Of Polymer Induced Macrophage

phage (4). This enzyme profile was used to evaluate the activation/differentiation state of our test polymer-elicited macrophage. Peritoneal exudate cells were collected from mice which had received eliciting agents 7 days prior to harvest. As had been previously reported (4), 5'N was found in significant quantities only in resident populations while APD is clearly a marker for thioglycollate/inflammatory macrophages. The LAP levels did not clearly discriminate between any of the populations although it had previously been reported to be elevated in "activated" PEC populations (4). The pattern of enzyme activities in the profile was most discriminatory in 48-hour cultures; therefore this time point was used to evaluate the activities of test polymer-induced PEC, Table I. The 5'N activity is elevated in all our test polymer-elicited populations compared to thioglycollate, pyran or C. parvum-elicited PEC. Indeed, the levels detected in CDA-MA and MP-MA induced PEC which also possess enhanced tumor cytotoxicity approach the level of resident populations. It may be concluded that the cell populations which are activated to tumoricidal capacity by test polymers do not possess the low 5'N phenotype of conventional activating agents. Test polymer-elicited PEC do not share the elevated APD phenotype of inflammatory populations. The LAP activities, once again, were not enhanced in "activated" populations. The ectoenzyme profiles of test polymer-elicited PEC populations assayed immediately after adherence (0 hour cultures) demonstrated similar patterns of activity.

Effect of Polyanionic Compounds on Growth of Lewis Lung and Mouse
Survival in Vivo

Mice were injected with 2×10^4 Lewis lung tumor cells in

Table II: Enhanced Survival of Test Polymer Treated Mice

Dosage (mg/kg)	25	50	160
Agent			
CDA-MA	>130%	>160%	>173%
MP-MA	95%	115%	>158%
IAS	>130%	>120%	> 80%
LS-MA	102%	107%	98%
PYRAN		>142%	

the left flank on day 0. Test polymers and controls were administered i.p. on days 0, 1, and 2. Measurement of tumor growth was initiated when tumors become palpable (day 17) and twice weekly thereafter. Table II illustrates the enhancement of survival in mice which had received tumor and test polymers, C. parvum, pyran, or nothing. CDA-MA at 100 mg/kg and 50 mg/kg enhanced survival time 73% and 60%, respectively, above the tumor control. MP-MA (100 mg/kg) showed 58% enhancement. Pyran demonstrates 42% enhancement and C. parvum 30%, IAS and LS were toxic to mice at a dosage of 100 mg/kg. Only two of the IAS-treated mice and three of the LS-treated mice at this dosage were alive on day 3. CDA-MA (25 mg/kg), MP-MA (50 mg/kg) and IAS (UM-2) (25 and 50 mg/kg) had slight enhancing effects on mouse survival but were not more effective than pyran or C. parvum. Similar patterns of protection are seen in the tumor growth experiment. CDA-MA (50 and 100 mg/kg), MP/MA (100 mg/kg) and IAS (25 and 50 mg/kg) reduced tumor growth compared to animals which had received tumor alone. The remaining mice in the IAS and LS groups at 100 mg/kg show marked reduction of tumor growth.

Acknowledgements: Authors wish to acknowledge NIH grant AI-15612-03 for this support of this study. They also wish to acknowledge the contributions by Dr. Culbertson, Dr. Tirrell, H. Wilkins, J. Jones, D. Abayasekara, and C. Chung.

REFERENCES

1. Ottenbrite, R. M., Donaruma, L. G., Vogl, O., "Anionic Polymer Drugs", John Wiley and Sons, N.Y., 1980.
2. Baird, L. G. and Kaplan, A. M.: In, Anionic Polymeric Drugs, Donaruma, L. G., Ottenbrite, R. M. and Vogl, O., Ed., John Wiley & Sons, N.Y., 1980 (Ch. 5).
3. ibid, p. 186-187.
4. Morahan, P. S., Edelson, P. J. and Gass, K.: J. Immunol., 125: 1312, 1980.

A RESPONSIVE HYDROGEL AS A MEANS OF PREVENTING CALCIFICATION IN UROLOGICAL PROSTHESES

E. C. Eckstein, L. Pinchuk* and M. R. Van De Mark

Departments of Biomedical Engineering and Chemistry
University of Miami, Coral Gables, Fla. 33124

INTRODUCTION

Hydrogel prostheses made of poly(2-hydroxyethyl methacrylate), pHEMA, have demonstrated resistance to calcification in the urinary tract(1-3). Implantation studies in non-urinary locations suggest that the methacrylic acid, MAA, content in these gels plays a critical role in their calcification resistance(4,5) and hence may be important for urine compatibility. It is well known that the equilibrium water content and other properties of the gel change significantly according to MAA content(6). Our work, presented here and elsewhere(7,8), shows that gels fabricated with small amounts of MAA and with low crosslinker content demonstrate a reversible shrink/swell behavior. The swollen volume or water content of these gels is highly dependent on electrolyte, pH, or urea contents of the test solutions. In response to a small change in the local environment the shrink/swell behavior may cause volume changes as large as seven hundred percent (largest swollen volume/smallest swollen value) and can occur quite rapidly for thin geometries. The responsive shrink/swell behavior occurs within the normal range of urine composition.

Recent research on polymer collapse by Tanaka et al.(9,10) shows that gel volume is the result of three constituent forces that set the osmotic pressure within the gel. These forces are rubber-like elasticity, polymer-polymer affinity and electrostatic interactions. Tanaka et al. used polyacrylamide gels in their research. The ionizable species, MAA, was incorporated into the polymer structure by hydrolyzing the polymerized gel. These gels were typically fabricated with 5% solids content. In solutions of low pH, the

----

* Current address: Cordis Res. Corp., Box 025700, Miami, FL 33102

323

hydrogen ion remains bound to the acid moiety on MAA and the gel
remains collapsed. As the alkalinity of the test solution is
increased, the proton dissociates, resulting in increased electrostatic repulsion and a swollen gel. Tanaka quantifies the electrostatic interactions through the concept of "hydrogen-ion pressure"
in which the dissociated protons are envisaged as a gas that in-
flates the gel(9,10). Greater amounts of ionization produce larger
pressures for swelling the gel. Electrolytes, e.g. calcium, sodium,
and potassium ions, shield the ionic charge on MAA and tend to
collapse the swollen gel. The polymer-polymer affinity character-
izes the preference of polymer chains to be near themselves instead
of near the solvent. The extents of swelling and collapse are
limited by the rubber-like elasticity of the molecular lattice(9,10).
(While hydrogels are not generally perceived as elastomeric materials, it is necessary to consider the entropic elastic effects of
the swelling process in the manner developed for rubbers and other
elastomers.) The degree of crosslinking in the gel and the molecular weight of the polymer are largely responsible for limiting the
swollen volume of the polymer lattice. The equilibrium state of the
gel represents a balance point among the rubber-like elasticity,
effects of the polymer-polymer affinity, and electrostatic interaction.

Of particular significance is the observation by Tanaka et al.
that in test solutions of mixed solvents (acetone and water), the
swollen volume has a pH or temperature dependence that mimics the
critical point behavior of phase transitions(9,10). Just as water
changes between liquid and vapor at 100 °C and 1 atm. of pressure,
the swollen volume exhibits a discontinuous change of value at a
critical value of pH or temperature. The corresponding experimental
observation is that a minute change of solvent condition causes a
huge change of swollen volume, either from a hyperswollen state to
a hyposwollen state or vice versa. (The term polymer collapse
originated because experimental studies showed that a large polymer
sample can become a much smaller sample for a minute change of
environmental condition.) In simple solvents, such as water, there
may still be dramatic steeply sloped change of swollen volume about
the critical point; however, a finite change of the environment
(e.g. a portion of a pH unit or a few °C) is necessary to change
between the hyperswollen and hyposwollen states. Discrete volume
changes at the critical point occur because unstable balances are
possible among the three forces(9).

The events noted above pertain to the equilibrium volume of the
gel and are not indicative of the time required to reach equilibrium
swelling. As volume changes are tied to diffusive rearrangement of
the polymer structure, changes of state can require extensive
lengths of time for bulk gels to reach equilibrium(9,10). However,
for surface zones or for thin grafted layers less material must
rearrange and the time required for swelling transitions may be

quite short. As details of the equilibrium swelling behavior depend upon the ionization, degree of crosslinking of the polymer, and the environment in which the gel is located, it is reasonable to expect that these factors also contribute to the kinetics of swelling.

Consideration of the biochemical environment of the urinary tract indicates that there are potent events that could change the magnitude of the polymer-polymer affinity and the electrostatic interaction in ionic gels such as MAA-doped pHEMA. The changes may cause the polymer to undergo a drastic change in volume. The composition of urine changes continuously depending upon food intake and metabolic activity of the individual(11). Electrolytes and urea form a large fraction of the solute content in urine. Urea, the end product of protein catabolism, is a potent swelling agent for hydrogels(12,13); in addition, it is a good solubilizing agent for calcium(14). Also, the pH of urine varies between 4 and 9 (11), and most importantly the pKa of MAA is within this range. These facts led us to expect that hydrogels containing the appropriate formulation of ionizable species and crosslinker could demonstrate an equilibrium swollen volume in slightly basic solution that is several times larger than that in moderately acidic solution. These responsive changes occur within the narrow pH range of urine composition.

METHODS AND RESULTS

The gels described here are based on 2-hydroxyethyl methacrylate, HEMA, with MAA added as an ionizable co-monomer prior to polymerization; solids content in the swollen gels are typically 20% to 60%. The crosslinker used was tetraethylene glycol dimethacrylate (TEGDMA).

For accurate study of the shrink/swell behavior of pHEMA based gels in various solvents, the purity of the starting materials must be known. HEMA monomer normally contains small amounts of impurities of MAA and crosslinker, ethylene glycol dimethacrylate (EGDMA). These impurities are byproducts from the synthesis of the monomer and/or from subsequent hydrolysis or transesterification of the monomer during transportation and storage(15). Purification steps, described elsewhere(7), are performed on all HEMA monomer batches prior to use. Samples of the purified HEMA lots are analyzed by HPLC to determine their purity. Any MAA or EGDMA remaining in the monomer (usually less than 0.01 mole percent) is quantified. Gels containing known amounts of MAA and crosslinker (TEGDMA) are made by doping the purified HEMA monomer with known amounts of the above comonomers and then polymerized with the appropriate initiators and solvents. Bulk gels are polymerized at 60% solids content following the chemical technique described by Ratner(13); our previously described pressurized polymerization technique is used to assure bubble-free gels(16). Grafted gel surfaces on polyurethane are made by the ceric ion technique described by Halpern(17). Once the purity of the HEMA monomer is established, gels displaying reproducible behavior are readily made. Samples of bulk gel for determination of equilibrium water content or swollen volume are made by cutting disks from a rod of hydrogel cast in a ten ml polypropylene syringe.

The swelling and shrinking behavior of the gels was explored by placing them in various swelling solutions simulating possible urological environments. The swelling environments, containing sodium chloride (9 g/l) and/or urea (9 g/l), were buffered to different pH values using buffering solutions described in Table 1. Baths of swelling solutions were at least 100 times the unswollen gel volume.

Table 1: Buffer compostions using 0.01 M solutions of the following acids and bases:

MEASURED pH	BUFFER COMPOSITION
2.1	HCl
2.8	Mixture of buffers at pH 2.1 and 4.0 *
4.0	Potassium acid phthalate **
4.8	Sodium acetate Acetic acid
5.8	Mixture of buffers at pH 4.8 and 7.0 *
7.0	Potassium phosphate monobasic ** Sodium hydroxide
8.1	Mixture of buffers at pH 7.0 and 9.2 *
9.2	Sodium tetraborate **
10.1	Sodium carbonate Sodium bicarbonate
11.0	Sodium hydroxide

* These solutions were made by adding the more basic or acidic buffer to the weaker buffer until the desired pH was obtained. Although these mixtures are not true buffers, continuous monitoring of the pH using a pH meter indicated that these solutions were essentially stable for the duration of the experiment.

** Buffers obtained from Fisher Scientific and diluted to 0.01 M.

The percent water content was calculated using the following equation:

% water content = (Ws - Wd) x 100/Ws

The swollen weight (Ws) was determined for each disk by first blotting the disk with a paper towel to remove excess water, and then weighing the swollen disk on an electronic balance. The dry weight (Wd) was measured after drying the sample in a convection oven overnight at 50°C. Measurements of gel size (diameter and thickness) were made using a vernier calipers; the volume was calculated using the formula for the volume of a cylinder. Three disks were equilibrated in each bath; in all cases the mean value of the mean value of the meausurements was calculated and presented.

Experiments show that bulk gels 2 to 3 mm thick may require 2 to 3 days at 40 °C to reach equilibrium. Many gels however, reach either the equilibrium swollen volume or a significant portion of it in several hours. For thin gels such as those found in hydrogel grafted surfaces (less than ten microns) the response to changes of local environment can occur within seconds to minutes. Further details of the transitional behavior are reported in (18).

Figure 1 shows the response of a pHEMA/MAA gel, doped with 2.5 and 0.15 mole percent of MAA and TEGDMA respectively, to its environment. Collapse of the gel appears as a dramatic change in volume, described by its water content versus pH. Gels that were equilibrated in the different pH solutions described above, in the absence of urea and salt, demonstrated critical-point-like behavior with the large change of volume occurring between pH 6 and pH 7. The swollen volume (80% water content) of the gel was approximately seven times its shrunken volume (40% water content). Addition of NaCl (9 g/l) to the buffered system lowered the maximum swelling to about 60% water content and decreased the slope of the swelling transition.

Realization of the degradation processes of urea is essential to interpreting the data on water content in solutions containing urea. The swelling of MAA-doped pHEMA gels in urea solutions with degradation is presented elsewhere(8). The fact relevant to these studies is that the degradation process produces ammonium hydroxide which acts to override the buffer solution. To insure that the buffers had not been exhausted, the pH of the buffer solution was measured at the time of assaying water content or swollen volume. The first observation is that addition of urea to the buffered pH solutions did not significantly alter the the degree of swelling at higher and lower pH values, that is at pH values away from the transition zone. Gels swollen in both urea and NaCl behaved much like those swollen in the salt by itself. This shows that urea by itself is not a potent swelling agent. If it were, the points for acidic solutions with urea would be raised above their counterparts

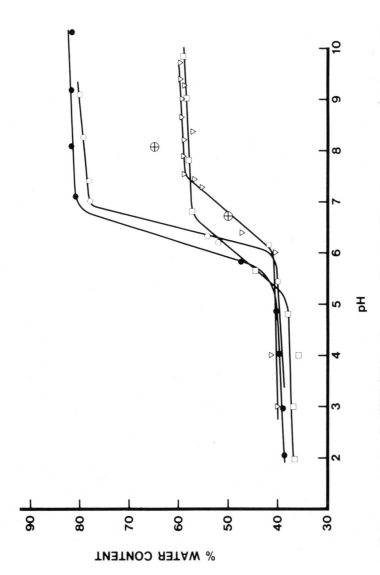

Figure 1: Water content of pHEMA/MAA gels vs. pH of buffered aqueous solutions. In addition to buffer described in Table 1, the solutions contain: ● no additional material, ○ urea (9 g/l), □ NaCl (9 g/l), ▽ NaCl and urea (both at 9 g/l). Two points ⊕ showing swelling in urine are also provided.

without urea; further data is shown elsewhere(8).

An interesting question in need of further study is whether the addition of urea or salt to the solutions causes the transition zone of critical-point-like behavior to move or broaden. For clear interpretation of results such studies should be made in a mixed solvent system similar to the acetone and water solutions used by Tanaka et al.(9,10). Such a system will distinctly show the location of the discontinutiy and allow the determination of whether a solute like urea affects the critical point by changing the location of the balance point of the three forces or merely alters the nature of the balance in the region of the critical point. A crucial fact in the presented data is that urea and saline do not move the responsive behavior out of the pH range of normal urine or reduce the scale of volume change to a trivial level.

Points showing the effects of urine at two different pH values on the swelling behavior of the gel are included in Figure 1. The point at pH 8.0 was adjusted to this value by addition of a small amount of sodium carbonate to the urine sample at pH 6.5. Quantitative comparison with the other curves is not possible since the osmolality and chemical composition of the urine are unknown. The 15% difference of water content of these gels in urine at different pH verifies the fact that gels can be made that are responsive to changes in urine composition. The difference in physical size is not obvious from the difference of water content; gels swollen in urine at pH 8.0 are twice as large as gels equilibrated in urine at pH 6.5. HEMA gels that are made without ionizable groups will not demonstrate this drastic swelling behavior(7) unless, of course, ionizable groups in the polymer are formed by subsequent hydrolysis, or other degradation processes peculiar to the gel.

DISCUSSION

Results indicate that MAA values as low as 1 mole percent with crosslinker concentration less than 0.3 mole percent provide sufficient internal ionization to allow great changes in volume to occur within the pH range normally found in urine (pH 5-9)(7). An acid moiety having a different pKa should move the collapse point; however, evidence that this point also depends upon the major monomeric constituent is given by the combination of our work and that of Tanaka et al. Their gels of MAA and acrylamide show collapse at pH 3.8 (9,10), which is considerably less than that observed for the pHEMA/MAA.

In light of the above results and because previous grafted prostheses used HEMA monomer as commercially supplied, the shrink/ swell behavior of retrieved implants was investigated. Short term tests (6 and 8 weeks) of prosthetic urinary bladders with surfaces grafted using unpurified HEMA monomer resisted calcium encrustation even in the presence of infected urine. Microscopic examination of the hydrogel surface after autopsy revealed a surface structure that was four times thicker than acid-shrunk or non-implanted controls. This swelling behavior could only have occured if the gel contained some finite amount of ionic species(7). HPLC analysis of residual stock of the HEMA monomer used for these implants indicate that impurities of MAA and EGDMA as high as 6% and 0.5% respectively, may have been present in those batches. Similar experiments were made on a retrieved ureteral implant of a bulk hydrogel, kindly supplied to us by Dr. Norman L. Block. The prosthetic ureter demonstrated shrink/swell behavior when cycled in acidic and basic solutions, again implying the presence of MAA or other ionizable species in the gel lattice. Hydrogel ureters of this nature remained patent in the canine urinary tract for periods of up to three years(1).

Several factors may influence the calcification resistance of hydrogels in the urinary tract; these factors are likely to differ in levels of importance from those for similar gels implanted in physiologically controlled sites, for example, studies of osteogenesis(4), or reconstructive surgery(19). In most of the body the environment of the gel is very stable and the physical and chemical inertness of the gel is a key factor. However, the ability of the gel to respond to local environment changes by shrinking or swelling may be related to events that improve their patency in the urinary tract. While the HEMA gels with ionizable content are somewhat chemically inert (e.g. hydrolysis is not a reported problem), they are not physically inert. The water content of pHEMA-based hydrogels gels is normally about 40%. As shown here the water content of pHEMA gels with minor amounts of MAA can range to 80%. This larger water content may enhance the action of the hydrogel as a biological interface, a concept elaborated by Andrade(20) and prevent precipitates from nucleating on the gel wall. The tensile strength of gels decreases as the water content increases(6). Krindel and Silberberg show that for certain flow rates, gel tubes can have a higher hydrodynamic resistance to flow than rigid tubes(21). Via dimensional analysis of the resistance data, they connect this behavior with the elasticity of the gel and hypothesize an irregular motion at the gel-liquid surface. Swollen gels with their lower moduli of elasticity, may allow ripples to form more readily on the surface than shrunken gels and thus may prevent calcium salt precipitates from adhering to the gel wall or dislodge small precipitates as they are formed. Another contributing mechanism may be that the swelling and shrinking motion of the gel itself physically sloughs calcium precipitates, if in fact, they form.

The nature of calcification resistance in the biological urinary tract is not well understood. However, it is interesting to note that the entire tract is lined with a layer of mucopolysaccharides, which are ionically charged gel-like substances containing large amounts of imbibed water. Structures made with pHEMA gels

330

interfacing urine may perform similar functions as these mucopoly-saccharides(22). In order to better understand the calcification mechanism and to design materials with long term viability in the urinary tract, the effects of ionizable groups in the gels and the subsequent swelling behavior of these gels is currently being ex-plored. Early data, from studies of grafted prosthetic ureters implanted in dogs for 6 months indicate the effectiveness of respon-sive surfaces. No encrustation has been observed at this time (18). However, longer term studies, which are in progress, are needed to prove the concept.

CONCLUSION

Hydrogels composed of pHEMA doped with small amounts of MAA and crosslinker demonstrate large volume changes between pH 6 and pH 7. The range of this collapse-like behavoir is situated well within the normal range of urinary pH, electrolyte and urea content. Calcifi-cation in the urinary tract is not well understood but resistance of HEMA gels to calcification could quite reasonably depend upon the "active" nature of the gel. Accordingly, the physicochemical be-havior of gels is as important to the design of urinary biomaterials as the simple chemical structure and physical properties.

ACKNOWLEDGEMENT

This work sponsored by NIH grant AM26630.

REFERENCES

1. N. L. Block, E. Stover, V. Politano, A Prosthetic Ureter in the Dog, Trans. Amer. Soc. Art. Int. Organs. 23:367 (1977).
2. S. Kocvara, C. Klement, J. Kubat, M. Stol, Z. Ott, J. Dvarak, Gel-Fabric Prostheses of the Ureter, J. Biomed. Mat. Res. 2:489 (1967).
3. B. Levowitz, J. LaGuerre, W. S. Calem, F. E. Gould, J. Scherrer, and H. Schoenfeld, Biologic Compatibility and Applications of Hydron, Trans. Amer. Soc. Art. Int. Organs, 14:82 (1968).
4. J. Smahel, J. Proserova and E. Behounkova, Tissue Reactions After Subcutaneous Implantation of Hydron Sponge, Acta Chirurgiae Plasticea, 13:193 (1971).
5. E. Cerny, R. Chromecek, A. Opletal, F. Papousek and J. Otoupa-lova, Tissue Reaction in Laboratory Animals to Some Varieties of Glycol Methacrylate Polymers, Scripta Medica 43:63 (1970)
6. M. Ilavsky, K. Dusek, J. Vacik, J.Kopecek, Deformational, Swelling, and Potentiometric Behavior of Ionized Gels of 2-Hydroxyethyl Methacrylate- Methacrylic Acid Copolymers, J. Appl. Polym. Sci. 23:2073 (1979).
7. L. Pinchuk, E. C. Eckstein and M. R. Van De Mark, Effects of Low Levels of Methacrylic Acid on the Swelling Behavior of Poly (2-Hydroxyethyl Methacrylate), J. Appl. Polym. Sci., In press.

8. L. Pinchuk, E. C. Eckstein, and M. R. Van De Mark, The Interaction of Urea with the Generic Class of Poly(2-hydroxyethyl methacrylate) Hydrogels, J. Biomed. Mater. Res., In press.
9. T. Tanaka, S-T Sun, I. Nishio. Phase Transition in Gels, In: "Scattering Techniques Applied To Supramolecular and Nonequilibrium Systems," S-H. Chen, B. Chu and R. Nossal, ed., Plenum, New York (1981).
10. T. Tanaka, Gels, Scientific American, 244, #1:124 (1981).
11. A. White, P. Handler, E.L. Smith, R. Hill, I.R. Lehman, "Principles of Biochemistry" McGraw-Hill, New York, (1978).
12. M. F. Refojo, Hydrophobic Interaction in Poly(2-hydroxyethyl methacrylate) Homogeneous Hydrogel, J. Polym. Sci. A-1 5:3103-3113 (1967).
13. B. Ratner, I. Miller, Interaction of Urea with Poly(2-hydroxyethyl methacrylate) Hydrogels, J. Polym. Sci. A-1 10:2425 (1972).
14. P. Boldrini, New Chemical Interpretation of What Calcifies in the Arteries, J. Theor. Biol. 95:325 (1982).
15. O. Wichterle, Hydrogels, In: "Encyclopedia of Polymer Science and Technology," 15:273, John Wiley and Sons, New York, (1971).
16. L. Pinchuk and E, C. Eckstein, Pressurized Polymerization for Reaction Casting of Poly(2-hydroxyethyl methacrylate), J. Biomed. Mater. Res. 15:183 (1981).
17. D. B. Halpern, O. Solomon, D.G. Chowhan, Polymer Studies Related to Prosthetic Cardiac Materials Which are Non-Clotting at a Blood Interface, Natl. Tech. Inform. Ser., #PB-255913/6WV (1976).
18. L. Pinchuk, Physicochemical Factors that Affect Copolymeric Gels of Poly(2-hydroxyethyl methacrylate / methylacrylic acid) as Materials for Urinary Tract Prostheses, Phd. Dissertation, University of Miami, submitted May, 1984
19. L. Sprincl, J. Vacik and J. Kopecek, Biological Tolerance of Ionogenic Hydrophilic Gels, J. Biomed. Mater. Res. 7:123 (1973).
20. J. D. Andrade, R.N. King and D.E. Gregonis, Probing The Hydrogel/ Water Interface, In: "Hydrogels for Medical and Related Applications," Ed. J.D. Andrade A.C.S. Symp. Series.#31, Washington, D. C. (1976).
21. P. Krindel and A. Silberberg, Flow Through Gel-Walled Tubes, J. Coll. Interface. Sci. 71:39 (1979)
22. C. L. Parsons, C. Stauffer and J. D. Schmidt, Bladder-Surface Glycosaminoglycans: An Efficient Mechanism of Environmental Adaptation, Science, 208:605 (1980).

POTASSIUM ION TRANSPORT THROUGH HYDROGEL MEMBRANES IN THE PRESENCE

OF BLOOD COMPONENTS: PLASMA PROTEINS

G.S. Margules, J.A. Kane, A.R. Livingston and
D.C. MacGregor
Cordis Research Corporation
P.O. Box 525700
Miami, Florida 33152-5700, U.S.A.

ABSTRACT

A mathematical model and conceptualization of membrane ion transport by diffusion and phase transfer as coupled to adsorbed protein associated ion specific adsorption has been developed. The model is verified by monitoring K^+ transport through poly-(2-hydroxyethyl methacrylate) membranes in the presence of plasma in concentrations of 20, 50, 80 and 100% by volume. The transport data can be adequately described by the model where the surface oriented K^+ is assumed to be constant and related to bulk protein content. Control studies provide a baseline value for an effective diffusion coefficient for K^+, D_{eff}, of 2.82 x 10^{-8} cm^2/sec. The addition of plasma to the transport study solutions shows an increase in the transport of K^+ such that the rate is exponentially related to plasma percent. Since the dimension of the albumin molecule is appropriate to reside in the surface convection layer (100 Å) and albumin itself is the majority protein in plasma, values of known albumin bulk concentrations and published surface concentrations are useful for extended analysis. The surface concentration of K^+ at 100% plasma is 12.9 nEq/cm^2 and the ratio of ion to protein is 4.3 Eq K^+/mM albumin. These results provide insight into the role of protein adsorption to biomaterials as they relate to ion transport. The transport chamber approach is ideally suited to modification of the biomaterial interfacial microenvironment where, as in this study, the effects of plasma protein interactions are separated from cellular interactions. In medical devices, this phenomenon is important in the design of electrochemical blood chemistry sensors and artificial kidney membranes. Key words: pHEMA, Protein Adsorption, Biomaterials, Ion Transport.

INTRODUCTION

Surface phenomena associated with biocompatibility are being studied from various viewpoints in an effort to understand, at a molecular level, the factors which relate largely to thrombogenicity. In many medical devices, such as blood chemistry sensors and artificial kidneys, the transport properties of the biomaterial are the key to their functionality. In these instances, the effects of protein adsorption on ion transport are conceptualized and verified to elucidate the role of protein associated ion specific adsorption as coupled to ion phase transfer and membrane mass transport.

Protein adsorption is thought to be the initial event which occurs when blood contacts a foreign material. The next events are platelet adhesion and fibrin formation which ultimately lead to blood coagulation. The adsorption of plasma proteins is therefore recognized as an important interaction in the thrombogenesis sequence. The formation of a proteinated surface on various polymers has been studied using radioactive labeling techniques. The proteination of polydimethylsiloxane, segmented polyether polyurethane, and copolymers of polyether urethane urea was characterized by Lyman (Lyman et al., 1974). Kim (Kim et al., 1974) studied protein adsorption as a precursor to platelet adhesion on polydimethylsiloxane, fluorinated ethylene/propylene copolymer, and a segmented copolyether urethane urea. In these studies, the plasma proteins albumin, globulins, and fibrinogen are shown to play the major role in subsequent thrombus formation. The polymer utilized in this study is poly(2-hydroxyethyl methacrylate) (pHEMA). Protein adsorption on similar polymers from a hemocompatibility standpoint, has also been studied by others (Horbett, T.A., 1981; Horbett, T.A. and Weathersby, P.K., 1981).

Plasma proteins are composed of amino acids with polar and nonpolar groups. The polar nature of proteins raises the possibility of ionoprotein complex bonding to high affinity binding sites on polymer surfaces.

These electrochemical events are particularly important in medical devices which rely on ion transport for proper operation. The primary examples are sensors which measure blood constituents via ion-selective electrodes (LeBlanc et al., 1976; Hill et al., 1978; Hill, 1981; Case et al., 1979) ISFET devices (McKinley et al., 1980) and coated wire technology (Cattrall et al., 1974).

Medical devices most relevant to this study utilize the same polymer (pHEMA) to enhance the biocompatibility of ISFET devices (Shimada et al., 1980), intravascular Clark-type oxygen electrodes (Gold et al., 1975), in vivo reference electrodes (Margules et al., 1982), and as an ion exchange membrane for an artificial kidney (Akizawa, T. et al., 1981). Here, the protein adsorption, as coupled to ion transport, is the primary phenomenon associated with accurate and reliable operation. Another area, which depends on the knowledge of how protein adsorption affects transport, is the controlled release of bioactive materials (Langer et al., 1981). pHEMA is commonly employed in this situation as well. Furthermore, these interfacial electrochemical events may be important in biological systems where specific adsorption of ions to cell membranes may control cellular processes (Pilla et al., 1977).

THEORY

The electrochemical events which occur at the pHEMA/electrolyte interface which are of primary interest within the framework of biocompatibility are specific adsorption and phase transfer. Here, the specific adsorption is viewed as ions competing with water dipoles for sites within the electrical double layer or ions associated with the polar amino acids of proteins which have preferentially adsorbed to the polymer surface. These types of interfacial interactions are usually confined to within 100 Å of the surface. Phase transfer is an energetic step where an ion transfers from the aqueous phase to the membrane phase. Once in the membrane phase, the ion is no longer influenced by surface phenomena and can transport through the membrane by diffusion or migration. The double layer formation is electrostatic in nature. Aqueous diffusion is only present in low molarity electrolyte solutions, thus minimizing the effects of these interactions on the transport of ions.

The overall transport process can then be viewed as a two-step process where the ion reaches the surface by simple diffusion, specifically adsorbs and then undergoes phase transfer. Any surface interaction which may modify the surface concentration of ions could affect the number of ions available for phase transfer. The transport of ions through pHEMA has been shown to be dominated by concentration gradient diffusion which follows Fick's first law of steady state diffusion (Chen, 1979).

The permeation of water in pHEMA has been characterized (Wisniewski et al., 1980) and the solute transport has been described (Lee et al., 1978).

To study the effects of protein adsorption on ion transport through hydrogel membranes, the overall process is described analytically in Equation 1.

$$\frac{V_2 dC_2}{dt} = -D_{eff}A_m \left(\frac{dC}{dx} + \frac{v\Gamma_\rho}{\delta_A} \right)$$

(1)

Equation 1 is a mass balance of a two-compartment system where V_2 is the volume of compartment 2, C_2 is the concentration in compartment 2, D_{eff} is the effective diffusion coefficient, A_m is the membrane surface area, C is the variable concentration, v is an effective coverage coefficient, Γ_ρ is the bulk concentration of protein and δ_A is the thickness of the surface convection layer. In this case, the units of v are represented by the surface concentration of ions/bulk concentration of protein. From Equation 1 the assumption that the proteination of the surface is a rapid process and reaches a steady state equilibrium is apparent. Equation 1 can be solved analytically and expressed in a form which is ideally suited for graphical solutions to verification experiments (Wisniewski et al., 1976).

$$\ln \left(C_1 - C_2 - \frac{v\Gamma_\rho \delta_m}{\delta_A} \right) = \ln (C_1 - C_{2i}) - \left(\frac{D_{eff}A_m}{V_2 \delta_m} \right) t$$

(2)

Equation 2 assumes steady state diffusion, steady state protein adsorption, the Goldman assumption for thin membranes, negligible migration, and zero protein adsorption at $t = 0$. In Equation 2, C_1 is the ion concentration in compartment 1, δ_m is the membrane thickness, and C_{2i} is the initial concentration in compartment 2.

MATERIALS AND METHODS

In an effort to quantitate the effects of protein adsorption on ion transport, polymer sheets are synthesized and transport chamber experiments are conducted.

pHEMA is polymerized using ammonium persulfate as the initiator, tetraethylene glycol dimethacrylate (TEGDMA, .15% by volume) as the cross-linking agent, and ethylene glycol as the solvent. Techniques similar to those developed by Pinchuk are utilized (Pinchuk et al., 1981). For assured homogeneity and

336

controllability large sheets are cast and cut into test discs. The discs are randomized and divided into control and experimental groups. The HEMA monomer (Sipomer® CL-100) was donated by Alcolac, Inc., Baltimore, MD.

The ion transport studies are conducted in a diffusion chamber consisting of two 10 ml acrylic compartments separated by a pHEMA test disc membrane where δ_m is equal to 1 mm. Saturated KCl is placed in compartment 1 and compartment 2 contains the experimental solutions. Chamber type experiments allow the simple modification of the microenvironment of one membrane surface by bulk solution variations. The control solution in compartment 2 is 4 mM KCl and the experimental solutions for compartment 2 are based on 4 mEq/l K^+ containing 20%, 50% or 80% human plasma by volume. The final experimental solution is 100% human plasma. In this case, K^+ transport from compartment 1 to 2 is monitored via an ion-selective electrode in compartment 2. A Beckman #39673 K^+ electrode, a Beckman quartz junction reference electrode (SCE) #B9 Model 39069 and a Beckman salt bridge #563853 are utilized in a standard manner with a Beckman Selection 2000 Ion Analyzer to measure K^+ concentration in compartment 2. The analog recorder output of the Beckman meter is connected to a Hewlett Packard logging multimeter #3467A which samples and prints data in a digital format as the natural logarithm of K^+ concentration in 10-minute intervals. The entire transport chamber and electrodes are immersed in distilled water at 37°C for accurate temperature control throughout each transport study. All experiments are conducted at 37°C with pHEMA test discs cut from the same sheet and with human plasma from the same initial supply. Randomized samples of plasma, in quantities of 50 ml each, are stored in a frozen state and handled the same way for each study. The experimental apparatus appears in Figure 1.

To solve Equation 2 graphically, the digital data are processed. The $\ln (C_1 - C_2)$ vs. t is plotted and subjected to linear regression analysis to obtain the slope and intercept. Here, the slope is $D_{eff}A_m/V_2\delta_m$ and the intercept is $\ln (C_1 - C_{2i})$. The value of $\ln [C_1 - \exp (\text{Beckman output, mV})]$ for each time point is computed on an HP85 desk-top computer utilizing an interactive basic program that is a modified HP standard pack "curve fitting program". Modifications allow keyboard entry of trial number, concentration, and number of samples. The calculations are done to seven significant figures and the program provides a listing of time, mV, $\ln (\Delta C)$, a plot of $\ln (\Delta C)$ vs. t, and a linear regression of the data. A sample appears in Figure 2. Similar transport studies are described previously (Margules, et al., 1982). The concentration of K^+ in compartment 1 is held constant (saturated) during the transport studies by adding an adequate amount of KCl crystals to the solution at time zero.

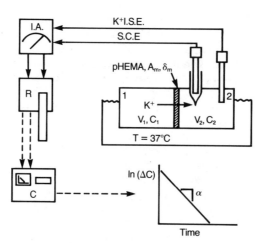

Figure 1. Transport experimental schematic. The chamber is labeled for compartment 1, compartment 2, volume 1 (V_1), volume 2 (V_2), concentration 1 (C_1), concentration 2 (C_2), K^+ flux and pHEMA membrane with surface area (A_m) and thickness (δ_m). The instruments are an ion analyzer (I.A.), data logger (R) and computer (C). The temperature is 37°C.

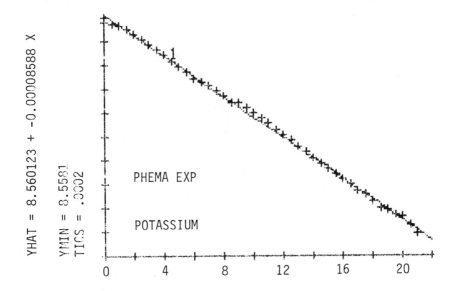

Figure 2. Sample of computer printout. The input is printed in the hours column and mV column. The computed ln (ΔC) is also printed for all times. The data are processed via least squares linear regression and the slope, intercept and correlation coefficient are printed. The data are then plotted and labeled.

RESULTS

The polymer synthesis results in homogeneous, bubble-free, and optically clear pHEMA sheets. The discs are stored in 4 mM KC1.

The transport studies provide both slope and intercept values for various compartment 2 test solutions. The results are summarized in Table 1. The intercept values are all the same since ln ($C_1 - C_{2i}$) is held at a constant at t = 0.

Table 1. Slope Data for Various Compartment 2 Solutions

Solution Compartment 2	(Slope/Hr)	Number of Experiments
4 mM KCl (control)	$7.77 \pm .42 \times 10^{-5}$	4
4 mM KCl, 20% plasma	$7.57 \pm .56 \times 10^{-5}$	4
4 mM KCl, 50% plasma	$7.72 \pm .35 \times 10^{-5}$	4
4 mM KCl, 80% plasma	$8.26 \pm .88 \times 10^{-5}$	4
100% plasma	$8.62 \pm .42 \times 10^{-5}$	4

ANALYSIS OF RESULTS

Figure 3 is a plot of the slope of ln (ΔC) vs. t for the different percentages of plasma in solution. From these data, it is possible to obtain a relationship for slope and % plasma. The best fit is obtained from a simple exponential where

$$\alpha = 7.53 \times 10^{-5} \exp [(1.15 \times 10^{-3}) (\%plasma)] \qquad (3)$$

The relationship in Equation (3) is used to compute the idealized values for slope (α_I). From the slope with only electrolyte in compartment 2, a value for D_{eff} is obtained using $A_m = .743$ cm^2, $\delta_m = .1$ cm and $V_2 = 10$ cm^3.

$$D_{eff} = \alpha_I V \delta_m /A_m = 2.82 \times 10^{-8} cm^2/sec \qquad (4)$$

This value is within the range of diffusion coefficients reported elsewhere (Migliaresi, C. et al., 1981).

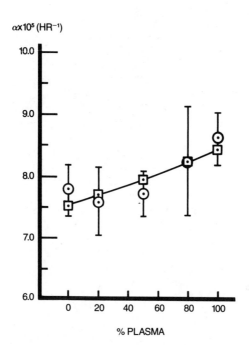

Figure 3. Plot of the slope of ln (ΔC) vs. t plots (α) vs. % plasma in compartment 2. The symbol ⊙ represents the experimental values shown with a single standard deviation and ▣ represents the idealized values for the best fit to a single exponential.

Since all of the transport studies result in plots similar to Figure 2 where the linear relationship holds for all time points (20-24 hours), the simple diffusion model with a constant adsorption term adequtely describes the transport phenomenon. Utilizing α_I for 0% plasma and assuming that D_{eff} in the membrane phase is independent of surface adsorption, a value for $v\Gamma_\rho \delta_m / \delta_A$ can be calculated from the α_I for 20, 50, 80 and 100% plasma.

To expand this analysis and obtain approximations for surface concentrations of K^+, albumin quantities (55%, 4.8 g/100 ml plasma, M.W. = 65,000) are utilized in the following computations. In this case, albumin is thought to be the majority species in protein adsorption to a foreign surface. At this point, Γ_ρ is the molar concentration of albumin in the bulk solution and δ_A is the convection layer thickness (100 Å). This is appropriate since albumin is approximately 160 Å in length. From the Γ_ρ and δ_A values, a value for v is straightforward. By utilizing the value of v, A_m and δ_A, it is possible to calculate the surface concentration of K^+ per unit area (Γ_K) or per unit volume (Γ_K').

The values of the above described parameters appear in Table 2. This analysis provides an insight into the surface oriented, protein associated K^+ content. Utilizing the information reported by Horbett (Horbett, 1981), where albumin was shown to adsorb to pHEMA significantly more than fibrinogen, immunoglobulin G, and hemoglobin at a surface albumin content of .08 $\mu g/cm^2$, it is possible to relate surface K^+ (Γ_K) to surface albumin. In the case of 100% plasma in the transport chamber, the ratio is 4.3 Eq of K^+ to every 1 mM of albumin.

DISCUSSION AND CONCLUSIONS

The proposed transport model couples both ion phase transfer and membrane mass transport phenomena to protein associated ion specific adsorption. The model views transport as a stepwise process where ions are first adsorbed and then undergo phase transfer. Potassium ion transport through pHEMA membranes is measured in the presence of 20, 50, 80 and 100% human plasma.

The adsorption/transport process can be adequately described by the model when protein adsorption increases the surface oriented potassium for subsequent phase transfer. The control studies resulted in a value of D_{eff} (K^+) of 2.82 x 10^{-8} cm2/sec. Increasing the plasma content of the experimental solutions resulted in an exponential increase in potassium transport. The adsorption process provides a constant surface concentration of K^+ since the transport data exhibits linearity over a 20 to 24 hour time range. Therefore, it is clear that plasma protein adsorption to pHEMA under these conditions enhances the transport of potassium ions.

342

Table 2. Analysis of Results, Slopes, Coverage Coefficients and Surface Concentrations

Solution in Compartment 2	α_I (HR^{-1})	$v\Gamma_\rho \delta m/\delta A$ mEq	Γ_ρ (Albumin) mM	v	ΓK nEq/cm^2	$\Gamma'_\rho K$ mEq/L
4 mM KCl	7.53 x 10^{-5}	0	0	0	0	0
4 + 20% plasma	7.70 x 10^{-5}	.18	.14	1.26x10^{-5}	2.4	2.4
4 + 50% "	7.97 x 10^{-5}	.46	.35	1.31x10^{-5}	6.2	6.2
4 + 80% "	8.25 x 10^{-5}	.75	.56	1.34x10^{-5}	10.1	10.1
100% "	8.45 x 10^{-5}	.96	.70	1.37x10^{-5}	12.9	12.9

343

Further analysis derived from known and reported albumin values provide an insight into the surface concentration of K^+ (Γ_K), and surface potassium/surface albumin ratio (Γ_K/Γ_A). The value of Γ_K at 100% plasma is 12.9 nEq/cm^2, and the Γ_K/Γ_A ratio is 4.3 Eq K^+/mM albumin.

In conclusion, these results elucidate the role of protein adsorption in overall transport processes. The phenomenon of enhanced transport is important to medical device applications in the areas of biomedical electroanalysis and artificial kidney membrane science. The transport chamber technique provides for the convenient modification of the interfacial microenvironment of biomaterials during transport studies. Here, the plasma component of blood was employed in an effort to separate the phenomenon associated with protein interactions from cellular interactions. This points to the flexibility of this procedure where the many environmental conditions associated with intravascular, heparinized body fluids or hemodialysis can be simulated in vitro.

ACKNOWLEDGMENT

The authors gratefully acknowledge the word processing contribution by D. Miller and B. Alcantara and the graphical art work by J. Orr.

REFERENCES

Akizawa, T., Koshikawa, S., Sasaki, K. and Nakabayashi, N., 1981, Development of ion-exchange membrane for an artificial kidney, Trans. Am. Soc. Artif. Intern. Organs., XXVII:644.

Case, R. B., Felix, A. and Wachter, M., 1979, Measurement of myocardial pCO with a microelectrode: Its relation to coronary sinus pCO , Am. J. Physiol., 236:H29.

Cattrall, R. W., Tribuzio, S. and Freiser, H., 1974, Potassium ion responsive coated wire electrode based on Valinomycin, Anal. Chem., 46:2223.

Chen, R. Y. S., 1979, Electrolyte transport through crosslinked poly-(2-Hydroxyethylmethacrylate), I. Effects of anionic size and crosslinking content, Polym. Prepr. Am. Chem. Soc. Div. Polym. Chem., 20:1005.

Gold, M. I., Diaz, P. M., Feingold, A., Duarte, I., Sohn, Y. J. and Kallos, T., 1975, A disposable in vivo oxygen electrode for the continuous measurement of arterial oxygen tension, Surgery, 2:245.

Hill, J. L., 1981, Intravascular K sensitive electrodes for clinical monitoring, in: "Progress in Enzyme and Ion Selective Electrodes," Lubbers, D. W., Acker, H., Buck, R. P., Eisenman, G., Kessler, M. and Simon, W., ed., Springer-Verlag, New York.

Hill, J. L., Gettes, L. S., Lynch, M. R. and Hebert, N. C., 1978, Flexible valinomycin electrodes for on-line determination of intravascular and myocardial K, Am. J. Physiol., 225:H455.

Horbett, T. A., 1981, Adsorption of proteins from plasma to a series of hydrophilic-hydrophobic copolymers. II. Compositional analysis with the prelabeled protein technique, J. Biomed, Mater. Res., 15:673.

Horbett, T. A. and Weathersby, P. K., 1981, Adsorption of proteins from plasma to a series of hydrophilic copolymers. I. Analysis with the in situ radio-iodination techniques, J. Biomed. Mater. Res., 15:403.

Langer, R., Brem, H. and Tapper, D., 1981, Biocompatibility of polymeric delivery systems for macromolecules, J. Biomed. Mater. Res., 15:267.

LeBlanc, O. H., Brown, J. F., Klebe, J. F., Niedrach, L. W., Slusarczuk, G. M. J. and Stoddard, W. H., 1976, Polymer membrane sensors for continuous intravascular monitoring of blood pH, J. App. Physiol., 40:644.

Lee, K. H., Jee, J. G., Jhon, M. S. and Ree, T., 1978, Solute transport through crosslinked poly(2-hydroxyethylmethacrylate) membrane, J. Bioeng., 2:269.

Lyman, D. J., Metcalf, L. C. Albo, D. Jr., Richards, K. F. and Lamb, J., 1974, The effect of chemical structure and surface properties of synthetic polymers on the coagulation of blood. III. In vivo adsorption of proteins on polymer surfaces, Trans. Amer. Soc. Artif. Int. Organs, XX:474.

Margules, G. S., Hunter, C. M. and MacGregor, D. C., 1982, A hydrogel based in vivo reference electrode, J. Electrochem. Soc., 129:135c.

Margules, G. S., Hunter, C. M. and MacGregor, D. C., 1982, A hydrogel based in vivo reference electrode catheter, Med. Biol. Eng. & Comp., 21:1.

McKinley, G. A., Saffle, J., Jordan, W. S., Janata, J., Moss, S. D. and Westerskow, D. R., 1980, In vivo continuous monitoring of K in animals using ISFET probes, Med. Instr., 14:93.

Migliaresi, C., Nicodemo, L., Nicolais, L. and Passerini, P., 1981, Physical characterization of microporous poly(2-hydroxyethyl-methacrylate) gels, J. Biomed. Mater. Res., 15:307.

Pilla, A. A. and Margules, G. S., 1977, Dynamic interfacial electrochemical phenomena at living cell membranes: Application to the toad urinary bladder membrane system, J. Electrochem. Soc., 124:1697.

Pinchuk, L. and Eckstein, E. C., 1981, Pressurized polymerization for reaction casting of poly (2-hydroxyethylmethacrylate), J. Biomed. Mater. Res., 15:183.

Shimada, K., Yano, M., Shibatani, K., Komoto, Y., Esashi, M. and Matsuo, T., 1980, Application of catheter-tip ISFET for continuous in vivo measurement, Med. Biol. Engr. & Comp., 18:741.

Wisniewski, S. J., Gregonis, P. E., Kim, S. W. and Andrade, J. D., 1976, Diffusion through hydrogel membranes. I. Permeation of water through poly(2-hydroxyethylmethacrylate) and related polymers, in: "Hydrogels for Medical and Related Applications," Andrade, J. D., ed., Am. Chem. Soc., Wash. D. C.

Wisniewski, S. J. and Kim, S. W., 1980, Permeation of water through poly(2-hydroxyethylmethacrylate) and related polymers: Temperature effects, J. Mem. Sci., 6:309.

INFLUENCE OF GEL AND SOLUTE STRUCTURE ON *IN VITRO*

AND *IN VIVO* RELEASE KINETICS FROM HYDROGELS

J.M. Wood, D. Attwood and J.H. Collett

Department of Pharmacy, University of Manchester

Manchester M13 9PL, U.K.

INTRODUCTION

Drug release kinetics from polymer implants *in vivo* are
influenced by the rate of drug diffusion within the polymer, and
by the nature of the medium surrounding the implant. Our previous
work has investigated the influence of gel structure on the
diffusion characteristics of solutes through poly(2-hydroxyethyl
methacrylate) (polyHEMA) hydrogels (1). The diffusion mechanism
was found to be influenced by the nature of the water within the
gel, and the average pore size of the network. Sorption of solutes
was reported to have a marked effect on network structure (2).

Diffusion through polymers is influenced not only by polymer
structure, but also by the molecular structure of the solute.
Solute structure may also influence the rate of partitioning into
the elution medium. However, there is little information on the
interrelationship between these factors. The present work is an
investigation of the *in vitro* and *in vivo* release kinetics of some
structurally related benzoic acids from both monolithic and
laminated polyHEMA gels. The influence of the physical structure
of the polymer network, the solubility, concentration and molecular
weight of the solute, and the presence or absence of a rate
controlling barrier at the surface of the matrix, have been
investigated.

MATERIALS AND METHODS

Materials

The following compounds were used as received: 2-hydroxyethyl methacrylate (HEMA) (98%) and ethylene glycol dimethacrylate (EGDMA) (99%), Fluka; benzoic acid (A.R.), Fisons Ltd.; 4-iodobenzoic acid, Koch-Light Laboratories Ltd.; 4-chlorobenzoic acid, 4-bromobenzoic acid, 4-hydroxybenzoic acid, 4-methoxybenzoic acid and 2-hydroxybenzoic acid (A.R.), B.D.H. Ltd.; 4-fluorobenzoic acid, Aldrich Chem. Co.; 4-toluic acid, Sigma Chem. Co.; ammonium persulphate, (A.R.), B.D.H. Ltd.; $7-C^{14}$ salicylic acid, N.E.N., specific activity 60 mCi $mmol^{-1}$.

Preparation of Hydrogels

a) *Chemically polymerized.* The aqueous monomer solution comprised 50% $^V/v$ HEMA, 0.2% $^W/v$ ammonium persulphate, and the required weight of solute (0-10% $^W/v$). The solution was deaerated by evacuation under a pressure of 50 mm Hg, bubbled with nitrogen, and poured into a mould comprising two glass plates separated by a spacing ring of height 0.5 cm. The solutions were incubated at 55°C for 16 hours. During polymerization the solute formed a homogeneous suspension in the gel.

b) *Radiation Polymerized.* Monomer solutions containing between 0 and 5% crosslinking agent (EGDMA) were prepared by a similar procedure to that described above, with the omission of the initiator. The solution was irradiated using a 2000 Ci Co^{60} gamma source with a dose of 300 krad at a dose rate of 5.79 krad min^{-1}. Radiation polymerized gels were used for all experiments except where indicated otherwise.

Preparation of Implants

Monolithic gels for implantation were prepared in cylindrical moulds 4.5 mm internal diameter and 1.0 - 1.5 cm in length. Implants contained 9.46 mg ml^{-1} of salicylic acid plus 2.6 - 3.0 μCi of $7-C^{14}$ labelled compound.

Laminated gels comprised a core containing 107 mg ml^{-1} salicylic acid and were coated with a crosslinked gel layer 1mm thick, containing no drug.

In Vitro Release Studies

Discs 2.35 cm diameter were cut from the gel and coated on the sides and base with silicone grease to prevent drug release from these surfaces. The discs were placed in the centre of a dissolution vessel (3) containing 300 ml of 0.0167 N HCl at 37°C.

This volume was chosen to ensure that sink conditions would prevail over the course of the experiment, for all the compounds under test.

The dissolution medium was stirred at 60 rpm and assay samples (2 - 3.5 ml) were removed periodically and replaced by the same volume of 0.0167 N HCl.

Assay Procedures

All samples were assayed by U.V. spectrophotometry (Pye Unicam SP500) after dilution in 0.1 N HCl. Values of λ_{max} were:- salicylic acid, 298 nm; 4-hydroxybenzoic acid, 255 nm; benzoic acid, 232.5 nm; 4-chlorobenzoic acid, 242 nm; 4-bromobenzoic acid, 246 nm; 4-iodobenzoic acid, 257.5 nm; 4-fluorobenzoic acid, 232.6 nm; 4-toluic acid, 240 nm; 4-methoxybenzoic acid, 257 nm. Interference from any unpolymerized HEMA (λ_{max} 212 nm) leached out from the discs during the course of dissolution experiments was shown to be negligible.

Determination of Solubilities

Accurate determination of solubility in the polymer matrix was not possible because of difficulties in direct assay of the compound in the gel. As an alternative, the solubility in the monomer solution was determined by spectrophotometric assay of solutions saturated at 37°C.

Determination of Partition Coefficients

Partition coefficients were determined by equilibrating known weights of gel (1.5 - 2.0 g) in 35 ml of 2.31×10^{-4} M solution at 37°C until there was no further change in concentration of the solution as determined by spectrophotometric measurement. Partition coefficients were expressed as the ratio of the concentration of solute present in the gel phase to that in the aqueous solution, at equilibrium.

In vivo Release Studies

Preliminary experiments were carried out to determine the route of excretion of the drug in rats. Following subcutaneous injection of aqueous solutions, over 99% of the administered dose was recovered in the urine within 18 hours. Excretion via other routes was therefore considered negligible.

Male Sprague-Dawley rats, 200 - 250 g were anaesthetised by intra-peritoneal injection of methohexitone sodium (50 mg kg^{-1}). Rods of gel were implanted subcutaneously in the back of the neck. Following implantation, the incision was sutured, and the animal placed in an all-glass metabolic cage, allowing separate collection of urine and faeces. Urine samples were analysed for C^{14} using a liquid scintillation counter (LKB Rackbeta 1216). Samples were mixed with a scintillation cocktail (Lumagel, Lumac) in the proportion 0.1 ml urine (diluted where appropriate) to 3 ml scintillant. The instrument corrected scintillation counts for quenching.

RESULTS AND DISCUSSION

Factors Influencing Release Kinetics *in vitro*

1. *Solute Structure*. Equations describing drug release rates from polymeric matrices of different geometries are well documented (4 - 9). For monolithic devices containing suspended drug, the amount of drug released under sink conditions is a linear function of the square root of time. The kinetics of drug release from a planar surface are described by (10):

$$Q = (D_m (2A - C_s) C_s t)^{\frac{1}{2}}$$ (1)

where Q is the amount of drug released after time t, D_m is the diffusion coefficient of the drug in the matrix, and A and C_s are the initial concentration and equilibrium solubility of the drug in the matrix respectively.
If $2A \gg C_s$, then

$$Q = (D_m 2AC_s t)^{\frac{1}{2}}$$ (2)

Fig. 1 shows the amount of solute (p-substituted benzoic acid) released *in vitro* from a planar surface of a monolithic gel, plotted as a function of root time. The linear relationship is indicative of a matrix controlled release mechanism (11). The slope gives the rate constant for release, which is dependent on D_m and C_s. Fig. 2 shows a linear relationship between the gradients of the plots of Fig. 1, divided by root D_m, against root solubility, in accordance with equation 2. However, benzoic acid gives a lower rate of release than would be predicted from values for D_m, as calculated from equation 1, and C_s, as determined experimentally. The low rate of release of this compound can be attributed to its high affinity for the gel matrix, as evidenced from its partition coefficient (Table 1). A high value of P may result in a partition-controlled release mechanism during the initial period of release if partitioning of the solute between the gel and the aqueous phase is slower than diffusion through the matrix. Chien and Lambert (11) have suggested that there may be a transition period between matrix controlled and partition controlled release for these solutes with high values of P. In the transition period, release is proportional to root time, but unlike a matrix controlled mechanism the slope of the $Q - t^{\frac{1}{2}}$ plot is influenced by P. Extrapolation of the plots for benzoic acid and I, Br and Cl-substituted compounds in Fig. 1

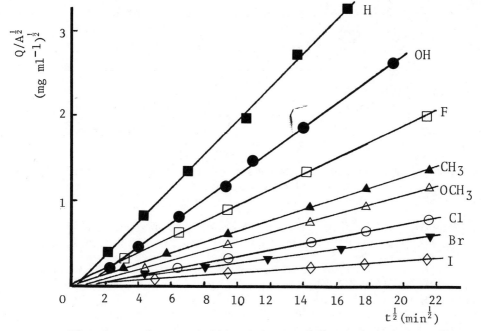

Fig. 1. Release profiles of p-substituted benzoic acids
from monoliths.

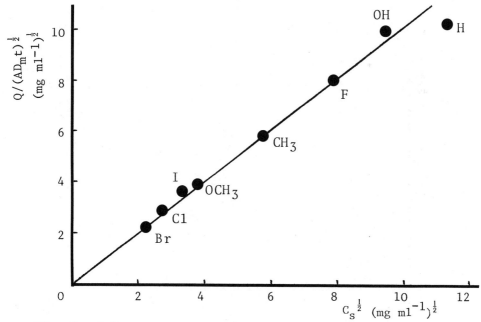

Fig. 2. Release rates of p-substituted benzoic acids as
a function of root solubility.

Table 1.   Partition Coefficients (P) of p-substituted (R) benzoic acids between polyHEMA
and water.

R	H	OH	F	CH$_3$	OCH$_3$	Cl	Br	I
P	127	13	2	14	18	25	36	75

indicates a lag time before the onset of steady state kinetics,
which may also be attributed to partition controlled release during
the initial period.

2. *Solute Molecular Weight.* The apparent diffusion coefficients, D_m, of the compounds, calculated from equation 1, are
plotted as a function of the cube root of the molecular weight in
Fig. 3, to allow a direct comparison of the effect of molecular size
and structure on the diffusion characteristics of the solutes.
A linear relationship is seen for five of the compounds, although the
low values of D_m for the –OH, –CH$_3$ and F substituted benzoic acids
(M = 136 – 140) suggests that D_m may also be dependent on other
factors for these compounds.   The linear relationship between
$Q/(AD_m t)^{\frac{1}{2}}$ and $C_s^{\frac{1}{2}}$ (Fig. 2) suggests that D_m is independent of A.

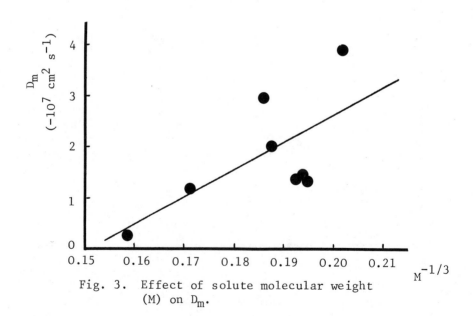

Fig. 3.   Effect of solute molecular weight
(M) on D_m.

Table 2. Slopes of Q vs $t^{\frac{1}{2}}$ plots for gel polymerized using a chemical initiator, and those polymerized by γ radiation.

Solute	Chemical (mg cm^{-2} min$^{-\frac{1}{2}}$. 10^2)	Radiation
4-H	45.98	58.18
4-OH	29.47	29.62
4-Cl	2.56	3.69
4-I	1.15	1.48

3. *Polymerization Method*. Table 2 compares the values of $Q/t^{\frac{1}{2}}$ for solutes in gels polymerized using a chemical initiator, with those polymerized by radiation. The slower release rates from chemically polymerized gels may be due either to a difference in gel structure resulting from the different methods of polymerization, or to differences in the particle sizes of the solute resulting from the slower polymerization (15 - 18h) of chemically polymerized gels. From a study of the influence of the polymerization rate on the permeability of aqueous polyacrylamide gels, Weiss et al (12) concluded that a more permeable gel is produced when the reaction rate is decreased, either by a reduction in temperature or by the use of lower initiator concentrations. Song (13) studied the influence of initiator concentration on the permeability of hydrogels to progesterone, and found that the permeability decreased as the initiator concentration increased. In view of these findings, gels polymerized by radiation might be expected to have lower permeabilities than chemically polymerized gels. The higher permeability noted in this study may be due to a more rapid precipitation of solute in the radiation polymerized gels causing the formation of smaller particles within the gel. As a consequence, C_s would effectively be increased and hence increase the release rate. The slower precipitation of solute in chemically polymerized gels may cause larger particles to form, the solubility of which is lower.

Factors Influencing Release Kinetics *in vivo*

Solute concentrations below saturation were used in monolithic devices implanted in rats, in order to ensure that absorption did not become the rate-limiting step in elimination, at high rates of release. Equations describing release kinetics of drugs from cylindrical devices are reported in ref (14). In the initial period, the amount of drug (Q_c) released at time t from a cylinder containing dissolved drug is given by:

$$Q_c/Q_\infty = 4(D_m t)^{\frac{1}{2}} r^{-1} \pi^{-\frac{1}{2}} - 3D_m tr^{-2} \tag{3}$$

where r is the radius of the cylinder and Q_∞ is the amount of drug released at infinite time.

For small values of t, the final term of equation 3 becomes very small, and release follows root time kinetics.

For large values of t, the amount of drug released is an exponential function of time.

The release kinetics during the initial period of release from a planar surface of a slab containing dissolved drug are described by (15):-

$$Q_s/Q_\infty = 2(D_m tL^{-2})^{\frac{1}{2}} \pi^{-\frac{1}{2}} \tag{4}$$

where L is the thickness of the sheet.

1.   *Crosslinker Concentration.* The effect of increasing the polymer crosslinker concentration on release rates both *in vitro* and *in vivo* was studied.   An increase in the crosslinking density (XL) of the polymer matrix reduces the solute diffusivity by reducing the porosity (ε), and increasing the tortuosity (τ) of the matrix (11), according to:

$$D_m = D \, \varepsilon/\tau$$

where D is the intrinsic diffusivity

$$\text{Thus } D_m = kD \, (XL)^{-1} \tag{5}$$

where k is a proportionality constant.

During the initial period of drug release, Q_c and Q_s are directly proportional to $D_m^{\frac{1}{2}}$, therefore

$$Q_c = k' \, (XL)^{-\frac{1}{2}} \tag{6}$$
$$\text{and } Q_s = k''(XL)^{-\frac{1}{2}} \tag{7}$$

where the value of the constant depends on the degree of hydration of the gel.

Plots of $Q_c t^{-\frac{1}{2}}$ and $Q_s t^{-\frac{1}{2}}$ against $(XL)^{-\frac{1}{2}}$ for benzoic and salicylic acids in gels of hydration 41% are shown in Fig. 4.   The gradients of the plots for the two compounds *in vitro* are very similar, indicating that at least for solutes of similar molecular structure, the nature of the solute does not affect the crosslinking of the polymer. However, a reduction in the hydration of the gel from 41% to 33% decreased the value of k' (Fig. 4) and further studies

354

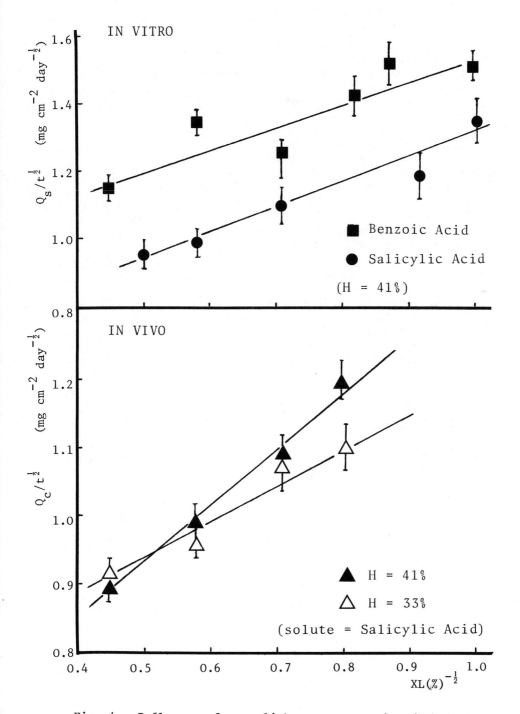

Fig. 4. Influence of crosslinker concentration (XL) on
release kinetics from gels *in vitro* and *in vivo*.

indicated that k' reduces to zero for a gel of hydration less than approximately 20%. This finding is in agreement with previous observations that in gels of low hydration diffusion occurs by solution diffusion, i.e. between the polymer chains, and not through the pores of the gel network (1). Therefore, it is to be expected that the crosslinker concentration will have little influence on drug release rates from gels of low hydration.

2. *Polymer Concentration.* The crosslinking density of the matrix depends not only on crosslinker concentration, but also on polymer concentration. The dependence of the crosslinking density of the network on the polymer concentration of polyHEMA gels has been demonstrated previously (16). The most rapid change in apparent crosslinking density with increase in polymer concentration occurred above a polymer concentration of 70% (Fig. 5), probably due to changes in the nature of the water within the gel. A similar dependence of D_m on polymer concentration has been noted for the diffusion of salicylic acid in polyHEMA (17).

Fig. 5 shows the influence of polymer concentration on *in vivo* release rates. It can be seen that the changes in release rate parallel the changes in crosslinking density, which is to be expected if release is matrix controlled. The results are of significance since they enable further predictions concerning drug release rates to be made using a knowledge of polymer structure. For example, the hydration of an implanted polymer gel may change slightly after implantation due to the diffusion into the gel of solutes in

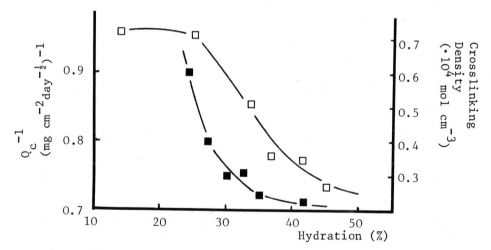

Fig. 5. Relationship between release rate (□) of salicylic acid from gels implanted *in vivo*, gel hydration, and polymer crosslinking density (■).

aqueous body fluids, causing the polymer to dehydrate or swell (2).
Thus it would be desirable to formulate the gel such that small
changes in hydration would not greatly affect the rate of release.
In the present system, this would indicate the use of gels of polymer
concentration greater than 70%.

     3. *Presence of a Rate Controlling Barrier.* The effect on
in vivo release rates of lamination of a rate controlling barrier on
to a monolith containing suspended drug is shown in Fig. 6.
Lamination produces a constant release rate which is desirable for
implanted dosage forms.

     The release rate of drugs from cylindrical laminated devices is
described by (18):

$$Q = 2\pi h \ r \ C_B \ D_B \ t \ 1^{-1} \tag{8}$$

where h is the height of the cylinder, r is the radius of the core,
C_B is the drug solubility in the barrier layer, D_B is the diffusion
coefficient and 1 is the thickness of the barrier layer. The
duration of the period of constant release of this low molecular
weight compound was relatively short (3 - 4 days). More lipophilic

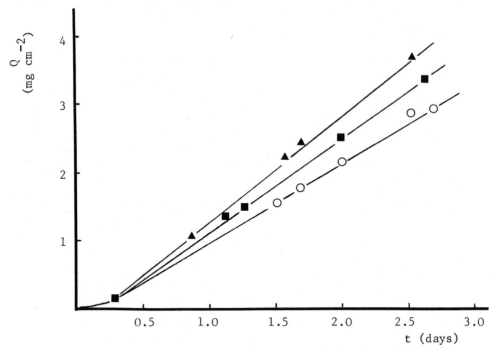

Fig. 6. Effect of crosslinker concentration in barrier
layer on release rates *in vivo* from laminates.
(▲) XL = 0.5%, (■) XL = 2%, (○) XL = 4%

drugs of higher molecular weight such as the steroids would have lower values of D_B and C_B and would be released over a much longer period.

Fig. 6 shows a reduction of release rate with increase in the crosslinking density of the laminated layer and suggests a method of control of release rate by this means.

In Vitro – *In Vivo* Correlations

The *in vivo* release rate of salicylic acid from a gel cylinder containing no added crosslinker was found to be 1.26 ± 0.06 mg cm^{-2} day $^{-\frac{1}{2}}$ for a gel of 41% hydration and initial solute concentration 9.46 mg ml^{-1}. The release rate *in vitro* from a planar surface of a gel of the same formulation was found to be 1.45 ± 0.102 mg cm^{-2} day$^{-\frac{1}{2}}$ The different thicknesses of the boundary diffusion layers *in vitro* and *in vivo* may account for some of the observed discrepancies. There was closer agreement between *in vitro* and *in vivo* release rates at higher crosslinker concentrations (Fig. 4) which may be due to a reduction in gel swelling with increased crosslinking. At a crosslinker concentration of 3%, the release rate *in vivo* ($Q_c/t^{\frac{1}{2}}$) was 0.99 ± 0.04 mg cm^{-2} day $^{-\frac{1}{2}}$, and the corresponding *in vitro* rate was 1.01 ± 0.08 mg cm^{-2} day $^{-\frac{1}{2}}$.

Other workers (19) found that *in vitro* release rates of ethnodiol diacetate from silicone devices depended on the stirring rate of the elution medium. Agreement with *in vivo* release rates hence depended on *in vitro* stirring conditions. Chien and Lau (20) observed faster rates of release of norgestomet from crosslinked polyHEMA gels *in vivo* than *in vitro*, while Roseman and Yalkowsky (21) postulated that release may follow a different kinetic order *in vivo* from that observed *in vitro* if the thickness of the diffusion layer was sufficiently large. Fibrotic encapsulation of the device *in vivo* hindering drug absorption may be another source of discrepancy between *in vitro* and *in vivo* release rates (22).

In summary, the results reported in this study indicate that when release is matrix controlled, estimates of drug release rates may be made from measurements of the physical and chemical crosslinking density of the polymer. Under specified conditions, the *in vitro* and *in vivo* release kinetics of salicylic acid were in good agreement, facilitating preliminary evaluation of dosage forms by *in vitro* tests alone. Zero order rates of release were achieved by lamination of a rate controlling barrier to the polymer, and the release rate modified through changes in the crosslinking density of the barrier layer.

REFERENCES

1. J.M. Wood, D. Attwood, and J.H. Collett, The Influence of Gel
   Formulation on Diffusion of Salicylic Acid in PolyHEMA
   Hydrogels. J. Pharm. Pharmacol. 34: 1 (1982)
2. J.M. Wood, D. Attwood, and J.H. Collett, The Swelling Properties
   of Poly(2-Hydroxyethyl Methacrylate) Hydrogels Polymerized by
   Gamma-Irradiation and Chemical Initiation. Int. J. Pharm.
   7: 189 (1981)
3. J.H. Collett, J.A. Rees, and N.A. Dickinson, Some Parameters
   Describing the Dissolution Rate of Salicylic Acid at Controlled
   pH. J. Pharm. Pharmacol. 24: 724 (1972)
4. H. Fessi, J.P. Marty, F. Puisieux, and J.T. Carstensen, Square
   Root of Time Dependence of Matrix Formulations with Low Drug
   Content. J. Pharm. Sci. 71: 749 (1982).
5. S.K. Chandrasekaran and R. Hillman, Heterogeneous Model of Drug
   Release from Polymeric Matrix. J. Pharm. Sci. 69: 1311 (1980)
6. R.H. Guy and J. Hadgraft, Theoretical Comparison of Release Rates
   of Drugs into Sink and Nonsink Conditions. J. Pharm. Sci.
   70: 1243 (1981)
7. T.J. Roseman and W.I. Higuchi, Release of Medroxyprogesterone
   Acetate from a Silicone Polymer. J. Pharm. Sci. 59: 353 (1970)
8. J. Cobby, M. Mayersohn, and G.C. Walker, Influence of Shape
   Factors on Kinetics of Drug Release from Matrix Tablets I:
   Theoretical. J. Pharm. Sci. 63: 725 (1974)
9. P.I. Lee, Diffusional Release of a Solute from a Polymeric
   Matrix - Approximate Analytical Solutions. J. Membr. Sci.
   7: 255 (1980)
10. T. Higuchi, Rate of Release of Medicaments from Ointment Bases
    Containing Drugs in Suspension. J. Pharm. Sci. 50: 874 (1961)
11. Y.W. Chien and H.J. Lambert, Differentiation between Partition -
    Controlled and Matrix - Controlled Drug Release Mechanisms.
    J. Pharm. Sci. 63: 515 (1974)
12. N. Weiss, T. van Vliet, and A. Silberberg, Influence of Poly-
    merization Initiation Rate on Permeability of Aqueous Poly-
    acrylamide Gels. J. Polym. Sci., Polym. Phys. Ed. 19: 1505
    (1981)
13. S.Z. Song, Hydrogel Devices for Controlled Drug Release, Ph.D.
    Thesis, University of Utah, (1980)
14. J. Hadgraft and R.H. Guy, Calculations of Drug Release Rates from
    Cylinders. Int. J. Pharm. 8: 159 (1981)
15. J. Hadgraft, Calculations of Drug Release Rates from Controlled
    Release Devices. The Slab. Int. J. Pharm. 2: 177 (1979)
16. J.M. Wood, D. Attwood, and J.H. Collett, Characterization of
    PolyHEMA Gels. Drug Dev. Ind. Pharm. 9: 93 (1983)
17. J.H. Collett, D. Attwood and J.M. Wood, Influence of Gel Structure
    on Diffusion through PolyHEMA Gels. J. Pharm. Pharmacol.
    33: 60P (1981)

18. E.S. Lee, S.W. Kim, S.H. Kim, J.R. Cardinal, and H. Jacobs, Drug Release from Hydrogel Devices with Rate-Controlling Barriers. J. Membr. Sci. 7: 293 (1980)

19. Y.W. Chien, S.E. Mares, J. Berg, S. Huber, H.J. Lambert and K.F. King, In Vitro – In Vivo Correlation for Intravaginal Release of Ethynodiol Diacetate from Silicone Devices in Rabbits. J. Pharm. Sci. 64: 1776 (1975)

20. Y.W. Chien and E.P.K. Lau, In Vitro – In Vivo Correlation of Subcutaneous Release of Norgestomet from Hydrophilic Implants. J. Pharm. Sci. 65: 488 (1976)

21. T.J. Roseman and S.H. Yalkowsky, Importance of Solute Partitioning on the Kinetics of Drug Release from Matrix Systems, in: "Controlled Release Polymeric Formulations", D.R. Paul and F.W. Harris, eds., A.C.S. Symp. Ser. 33,Washington D.C. (1976)

22. H.A. Nash, D.N. Robertson, A.J.M. Young, and L.E. Atkinson, Steroid Release from Silastic Capsules and Rods. Contraception 18: 367 (1978)

INTERACTION BETWEEN BLOOD COMPONENTS AND

HYDROGELS WITH POLY(OXYETHYLENE) CHAINS

S.Nagaoka, Y.Mori, H.Takiuchi, K.Yokota[*], H.Tanzawa,
and S.Nishiumi

Basic Research Laboratories, Toray Industries Inc.
and [*]Toray Research Center
1111 Tebiro, Kamakura 248, Japan

INTRODUCTION

When a biologically imcompatible material is in contact with
blood, there takes place thrombus formation on it via a rapid adsorption of plasma proteins and the subsequent adhesion of platelets.
Numerous approaches to supress the adhesion of blood components onto
synthetic surfaces have been studied and hydrogels are found to be
useful.

Recently, Abuchowski[1] has reported that long poly(oxyethylene)
(POE) chains covalently bound to bovine serum albumin significantly
reduce the immunogenicity of the bovine albumin in a rabbit, since
the flexible and hydrophilic POE shell with bound water surrounding
the bovine serum albumin covers the antigenic determinants. Taking
this excellent screening effect of POE chains from the immune process
into consideration, polymers with POE chains may be promising as a
blood-contacting materials which reduces the adhesion of the blood
components.

From this view point, we prepared hydrogels containing methoxy
poly(ethylene glycol) monomethacrylates with POE side chains of various chain lengths and investigated their interactions with blood components.

$$CH_2= \overset{\overset{\displaystyle CH_3}{|}}{\underset{\underset{\displaystyle CO-(\ OCH_2CH_2\ )_n OCH_3}{|}}{C}} \qquad (n=4,\ 9,\ 15,\ 23,\ 50,\ 100)$$

MATERIALS AND METHODS

Monomers

Methoxy poly(ethylene glycol) monomethacrylates (MnG) with the
POE chains of various chain lengths (n: degree of polymerization of
POE side chain) used in this study were obtained by the esterification of methacrylic acid with the methoxy poly(ethylene glycol) and
kindly supplied from Yushi Seihin Co., Ltd.

Synthesis of Random Copolymers

A mixture of MnG and methylmethacrylate (MMA) and a small amount
(0.05 mol%) of diethyleneglycol dimethacrylate (cross-linking agent)
was polymerized at 50 °C for 20 hours using azobisisobutyronitrile
as an initiator in a space between two glass plates. A cross-linked
copolymer was obtained in the form of a thin sheet and swollen in
water. The cross-linked copolymers of MnG and MMA are referred to
as P(MMA-co-MnG).

Synthesis of Graft Copolymers

Poly(vinyl chloride) (PVC) was reacted with sodium N,N-diethyl
dithiocarbamate in N,N-dimethylformamide (DMF) at 55 °C for 2 hours
to introduce the dithiocarbamate (DTC) groups, which are necessary
for the photo-induced graft copolymerization.

$$\{CH_2-\underset{\underset{Cl}{|}}{CH}\} + NaS-CSN(C_2H_5)_2 \xrightarrow[-NaCl]{} \{CH_2-\underset{\underset{SCSN(C_2H_5)_2}{|}}{CH}\}$$

Then MnG was graft copolymerized to the PVC containing the DTC
groups in tetrahydrofuran (THF) at 30 °C for 8 to 10 hours using a
100W high-pressure mercury lamp. Details of this polymerization was
reported elsewhere[2]. The reaction mixture was poured into methanol.
The precipitate was soaked in methanol for 2 days to remove the unreacted monomers and then dried under vacuum at room temperature.
The resultant polymers are referred to as P(VC-g-MnG). A sheet of
the P(VC-g-MnG) was prepared by repeatedly spreading the 5 wt% polymer solution in THF onto a glass plate and drying at 35 °C in a
nitrogen atmosphere.

Hydrogels

P(MMA-co-MnG) gels and P(VC-g-MnG) gels were obtained by washing the polymer sheets with methanol to remove the solvent or the
unreacted monomers and replacing methanol with water. The hydrogels
were preserved in distilled water. Hydrogels containing N-vinyl-
pyrrolidone (NVP) or glyceryl methacrylate (GLM) as a comonomer were
also prepared by the same method as mentioned above.

Characterization of Hydrogels

The water content (W) of the hydrogels, which is defined as follows, was regulated by changing the polymer composition.

$$W(\%) = \frac{\text{Weight of water in a gel}}{\text{Weight of a hydrated gel}} \times 100$$

^{13}C NMR spectra of the hydrogels were measured by a JEOL FX-100 pulsed FT spectrometer[3]. Free induction of decay signals ranging from 1000 to 12000 were accumulated at room temperature. Tetramethylsilane was used as an external reference for ^{13}C chemical shifts. The concentration of the ether carbon derived from POE chain existing on the surface of the polymers was measured by an electron spectroscopy for chemical analysis (ESCA) (KOKUSAI DENKI ES-200). The morphological structures of the polymers were observed by a transmission electron microscope (TEM) (JEM-100C). The ultra-thin sections of the specimen for the transmission electron microscopy were stained with osmium tetroxide (OsO4).

In Vitro Studies

The blood was withdrawn from the carotid of a rabbit into the siliconized glass tube containing a 3.8 wt% trisodium citrate aqueous solution. Plasma and platelet rich plasma (PRP) were prepared from the blood by centrifugation. The platelet counts (ca. 2×10^5 / μl) in the PRP were adjusted by diluting the PRP with the plasma. The hydrogels were soaked separately in the plasma or PRP at 37 °C for 3 hours. After rinsing the hydrogels several times with a saline solution, they were fixed with a 3 wt% glutaraldehyde saline solution. The amounts of plasma protein and total protein adsorbed by the hydrogels soaked in the plasma and PRP respectively were determined by an amino acid analyzer (HITACHI 835). Then the amount of adhered platelets was determined by subtracting the amount of adsorbed plasma protein from that of adsorbed total protein. The amount of adhered platelets was expressed as the relative value to the glass (Pylex 7740) used as a control in order to eliminate the individual variations of the experimental animals. The morphological deformation of the adhered platelets was observed by a scanning electron microscope (HITACHI-AKASHI MSM-4).

In Vivo Studies

Polyester sutures coated with the P(VC-g-MnG) and the P(VC-g-NVP) were implanted in the femoral vein of mongrel dogs for 2 hours, 1, 3, 7, 14 and 72 days. The sutures were evenly coated and the thickness of the coated layer was determined to be ca. 20 μm by a phase-contrast microscope (NIKON MD). The surfaces of the implanted specimens were observed by a scanning electron microscope (USM-50A).

The specimens for the scanning electron microscopy were fixed with a 2.5 wt% glutaraldehyde saline solution and stained with an aqueous 1 wt% osmium tetroxide solution at 4 °C.

RESULTS AND DISCUSSION

Properties of Hydrogels

Water content. The water content of the hydrogels in an equilibrium swollen state in water was dependent on the chemical composition of the copolymers and linearly increased with oxyethylene content in both the P(MMA-co-MnG) and P(VC-g-MnG) gels irrespective of the POE chain length (n). Figure 1 shows the relationship between the water content and oxyethylene content of the P(MMA-co-MnG) gels and as shown in Table 1, the water contents of the P(MMA-co-MnG) gels with similar oxyethylene contents of 38 - 40 wt% but various POE chain length (n) were as constant as 43 - 46 wt%.

Surface chemical composition. In designing a material for biomedical uses, it is of great importance to examine the surface characteristics which may strongly affect the biocompatibility of the material. From ESCA studies on the ether carbons of the surfaces of the freeze dried hydrogels listed in Table 1, it is demonstrated that the surface oxyethylene contents of these hydrogels were on the same level irrespective of the POE chain length (n).

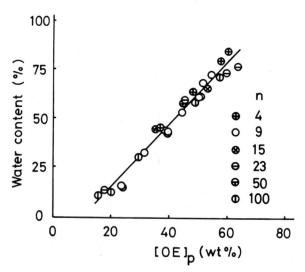

Figure 1. Relationship between water content and oxyethylene unit content ([OE]p) of P(MMA-co-MnG) gels.

Table 1. Chemical compositions of the typical P(MMA-co-MnG) gels.

Sample code	MMA (wt%)	MnG (wt%)	2G[a] (mol%)	[OE]$_p$[b] (wt%)	*C-O/C$_{1s}$[c]	W[d] (%)
P(MMA-co-M4G)	37	63	0.05	39.6	0.50	45.9
P(MMA-co-M9G)	50	50	0.05	39.4	0.46	43.8
P(MMA-co-M15G)	55	45	0.05	38.7	–	45.1
P(MMA-co-M23G)	56	44	0.05	40.1	0.47	43.4
P(MMA-co-M50G)	59	41	0.05	38.9	0.52	42.0
P(MMA-co-M100G)	60	40	0.05	39.0	0.49	44.7

[a] Diethylene glycol dimethacrylate.
[b] Content of oxyethylene unit in polymers.
[c] Ratio of ether carbon (*C-O) to total carbon (C$_{1s}$) measured by ESCA.
[d] Water content.

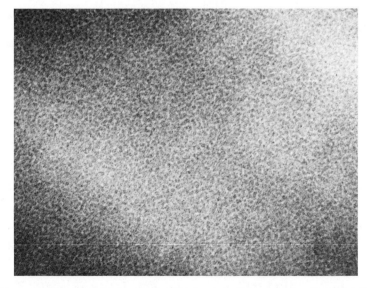

0.1 μm

Figure 2. Transmission electron micrograph of the ultra-thin section of P(VC-g-M9G) : [OE]$_p$ = 38.0 %, water content = 38 % ; stained with osmium tetroxide.

365

Surface morphology.   Block or graft copolymers composed of in-
compatible (immiscible) polymeric segments often form microphase
separated structures.   Figure 2 shows a transmission electron micrograph of an ultra-thin section of the graft copolymer, P(VC-g-M9G),
with the oxyethylene content of 35 wt%.   The spherical, dark microphase of 200 - 300 Å in diameter corresponds to the domain of POE
chains stained with OsO4 and the light, continuous phase corresponds
to the PVC part.   In case of the random copolymer, P(MMA-co-M9G),
with the oxyethylene content of 35 wt%, such a microphase separated
morphology was not observed.   Because short POE chains are distributed along the main chain.

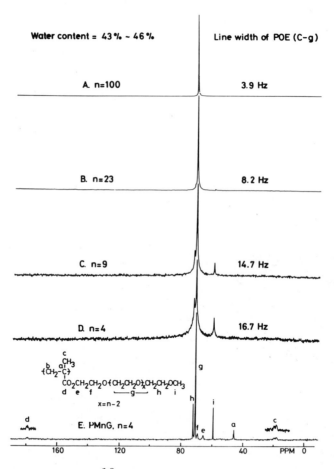

Figure 3.   25 MHz ^{13}C NMR spectra and line widths of P(MMA-
co-MnG) gels shown in Table 1 measured at 27 °C: A, n=100;
B, n=23; C, n=9; D, n=4.
The assignment of the signals was made on the basis of the
assignments for slightly cross-linked poly M4G gels.

Mobility of the polymer chain in the hydrogels. Proton decoupled ^{13}C NMR has been proved to be a very powerful tool for characterization of polymers including bulk materials such as solid rubbers above their glass transition temperature and synthetic or natural swollen gels. In these bulk materials, however, all the carbons in polymers do not contribute to the high resolution ^{13}C resonances, since a sizable loss of the peak areas occurs in case where immobilized segments such as cross-links or crystalline portions are contained. Figure 3 shows the ^{13}C NMR spectra of the P(MMA-co-MnG) gels with similar chemical compositions and water contents but various POE chain lengths listed in Table 1. In the ^{13}C NMR spectra of the P(MMA-co-MnG) gels at room temperature, only the ^{13}C resonances from the POE chains were observed. This can be explained in terms of the acquisition of the rapid motions of the POE chains as a result of swelling by water. The line width of the peak reflects the mobility of the POE chain and the narrower line width indicates the higher mobility. Though this is naturally dependent on the water content of the hydrogels, it was revealed from Figure 3 that the mobility of the POE chain increased with the increasing chain length in the P(MMA-co-MnG) gels having similar water contents.

Furthermore, POE chain showed higher mobility in the P(VC-g-MnG) gels than in the P(MMA-co-MnG) gels for the same POE chain length (n=9) as shown in Figure 4. This phenomenon suggests that the hydrated water molecules are more localized in the microdomain of the graft copolymers than in the homogeneous structure of the random copolymers.

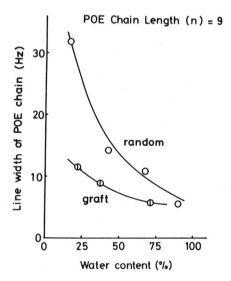

Figure 4. ^{13}C NMR line widths of POE chains of the random -type P(MMA-co-M9G) gels and graft-type P(VC-g-M9G) gels with various water contents.

367

In Vitro Studies

The effect of POE chain length (n). We investigated the effects
of the POE chain length (n) on adhesion of platelets and adsorption
of plasma protein obtained from rabbit, using the P(MMA-co-MnG) gels
with the similar water contents ranging from 43 to 46 wt%. As is
shown in Figure 5, the amounts of both the adhered platelets and the
adsorbed plasma protein significantly decreased with increasing POE
chain length (n) and as low as an almost negligible value and a value
below 0.1 μg/cm^2, respectively for n=100. As the adsorption and possible denaturation of plasma proteins on foreign surfaces has generally been considered as a trigger for the subsequent platelet adhesion,
a minimal protein interaction should lead to low platelet adhesion to
these hydrogels. In Figure 6 are shown scanning electron micrographs
of the surfaces of the hydrogels (n = 4, 23, 100) which were soaked
in the PRP at 37 °C for 3 hours. The amounts of platelets adhered
onto the P(MMA-co-NVP) and P(MMA-co-GLM) gels were even larger than
that onto the P(MMA-co-M4G) gels with the similar water content and
the shortest POE chain length. Figure 7 shows the scanning electron
micrographs of the platelets adhered onto the surfaces of P(MMA-co-
NVP) and P(MMA-co-GLM) gels with the water contents of 43 and 45 wt%
respectively. The adhered platelets were observed to deform flatly

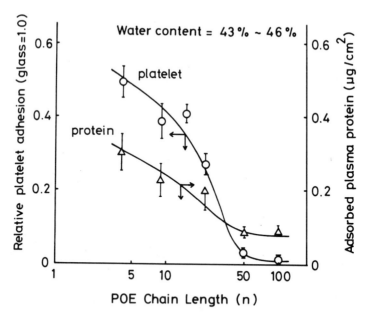

Figure 5. Effect of POE chain length (n) on the adhesion
of platelets and adsorption of plasma proteins onto the
P(MMA-co-MnG) gels shown in Table 1. Mean values and standard deviations (N=4) are shown.

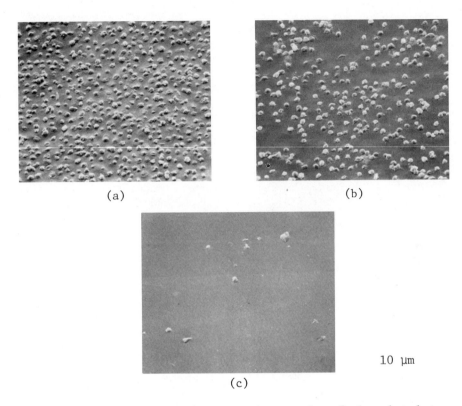

(a)   (b)

(c)

10 μm

Figure 6. Scanning electron micrographs of the platelets
adhered onto P(MMA-co-MnG) gels shown in Table 1:
(a), n=4; (b), n=23; (c), n=100.

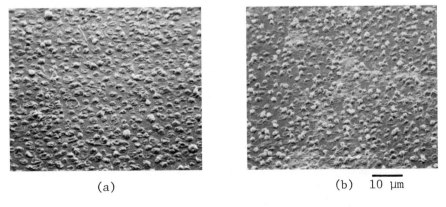

(a)   (b)   10 μm

Figure 7. Scanning electron micrographs of the platelets
adhered onto P(MMA-co-NVP) gel (water content = 43 wt%)
(a), and P(MMA-co-GLM) gel (water content = 45 wt%) (b).

and aggregate on these hydrogels. The value of the relative platelet adhesion onto these hydrogels was ca. 0.7.

To account for the observed effect of the long POE chain of the hydrogels, we propose the mechanisms schematically represented in Figure 8. Figure 8(a) indicates the volume restriction effect of POE chain. As many investigators dealing with polymeric dispersants have pointed out, irreversibly adsorbed nonionic macromolecules stabilize emulsions and suspensions in aqueous and non-aqueous media("protective effect"). The repulsive forces by which such adsorbed polymers can keep the particles dispersed in solution by counteracting van der Waals forces of attraction, are generated by the loss of possible chain conformations, as the volume available to the adsorbed chains is reduced between approaching surfaces. As Hesslink et al.[4] and Meier[5] have pointed out, the repulsive force generated by the flexible "tails" (chains terminally adsorbed at one end group) increases

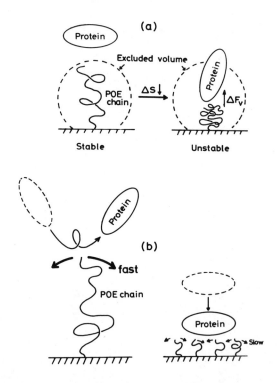

Figure 8. Schematic representation of the interaction between blood components and hydrated POE chains on a surface: According to the thermodynamical volume restriction effect (a) or the high mobility (b) of the hydrated long POE chain, blood components such as plasma proteins could not approach the polymer surface.

370

with the length and the amount of the polymer adsorbed per unit area. The hydrated, long POE chains at the surface of the hydrogel may exclude other hydrated biological polymers (e.g., plasma proteins or adhesive polysaccharide on the cell membrane of platelets) due to the mechanism mentioned above. Maroudas[6] have qualitatively pointed out this hypothesis.

Further , Figure 8(b) indicates the effect of the mobility of the POE chains. For irreversible adhesion, blood components have to be in contact with the foreign surface more than a certain measure of time. The rapid movements of the hydrated chains may prevent stagnation of the blood components on the surface of the hydrogels, probably because the contact time is shortened. The mobility of the hydrated POE chains increases with their chain length as shown in Figure 3, hence the long POE chains suppress the adhesion of the blood components more effectively than the short chains. The favorable behavior of the long POE chains could be explained by these mechanisms.

Effect of the morphological structures. The adhesion of the platelets and the adsorption of the plasma protein to the P(VC-g-MnG) gels of the graft-type were much weaker than those to the P(MMA-co-MnG) gels of the random-type having the corresponding POE chain length and the similar water contents, though PVC and PMMA in themselves adsorb the similar amount of blood components in vitro. Figure 9 shows the comparison of the amount of platelets adhered to P(VC-g-M9G) gels and P(MMA-co-M9G) gels with the same POE chain length and various water contents. These findings can be correlated with the microphase separated structure of the P(VC-g-M9G) gels shown in Figure 2. The adhesion of blood components to the hydrophilic microdomain may be markedly suppressed because both the local concentration and the mobility of POE chains are higher in this domain than in the corresponding homogeneous P(MMA-co-M9G) gels. Furthermore, as is obvious from Figure 2, the hydrophobic PVC domain is 100 to 200 Å in width and may not be wide enough to adsorb relatively large plasma proteins such as γ-globulin or fibrinogen which may induce the adhesion of platelets. For these reasons, the graft-type hydrogels show less platelet adhesion than the random-type ones.

In Vivo Studies

The polyester sutures coated with P(VC-g-NVP) (water content: 63.2 wt%), P(VC-g-M9G) (water content: 63.7 wt%), P(VC-g-M23G) (water content: 64.9 wt%), and P(VC-g-M100G) (water content: 69.1 wt%) were implanted in peripheral canine veins[7] for up to 72 days. High and moderate thrombus formations were found on the surface of the P(VC-g-NVP) and P(VC-g-M23G) respectively, whereas no thrombus formation was seen on the surface of the P(VC-g-M100G) even after a 72 day implantation. The scanning electron micrograph of the P(VC-g-M100G) implanted for 72 days (Figure 9) showed no adhesion of blood cells, although there was deposition of proteinous substances. More details

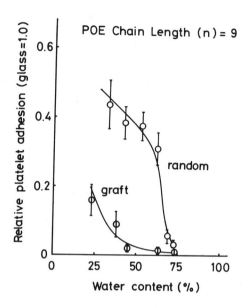

Figure 9. The amount of platelets adhered onto the random-type P(MMA-co-M9G) gels and graft-type P(VC-g-M9G) gels with various water contents. Mean values and standard deviations (N=4) are shown.

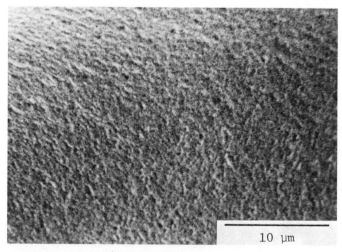

Figure 10. Scanning electron micrograph of the surface of the P(VC-g-M100G) implanted in the femoral vein of a dog for 72 days. No adhesion of the blood cells such as platelets and leucocytes is observed.

of these studies were reported elsewhere[8]. these findings are closely related to the in vitro studies described above.

Further, another ex vivo experiment using arteriovenous shunt of rabbit indicated that the hydrogels with long POE chain did not damage the function of the circulating platelets also (e.g., platelet count, adhesiveness and clot retraction ability) even when they contacted with the wide area of the surface[9]. This fact could be explained by the low amount of platelets adhered onto the hydrogels, because released substances from adhered platelets (e.g., ADP) suppress the function of the circulating platelets[10].

CONCLUSION

A new antithrombogenic material with long poly(oxyethylene) chains has been developed. The hydrated long poly(oxyethylene) chains existing on the surface of the hydrogels prevent the adsorption of blood components by their high mobility and volume restriction effect. The application of these polymers to blood-contacting materials of various artificial organs and medical supplies seems fruitful and is now under study.

ACKNOWLEDGEMENT

The authors are also grateful to Dr. Ishitani of Toray Research Center for ESCA measurements of polymers and to Dr. Noishiki for in vivo studies.

REFERENCES

1. A. Abuchowski, T. van Es, N. C. Palczuk, and F. F. Davis, Alteration of immunological properties of bovine serum albumin by covalent attachment of polyethylene glycol, J. Biol. Chem., 252:3578 (1977).
2. S. Nagaoka, M. Shiota, Y. Mori, and T. Kikuchi, Kinetics of photo-induced graft copolymerization to poly(vinyl chloride) containing diethyldithiocarbamate groups, Kobunshi Ronbunshu, 38:571 (1981).
3. K. Yokota, A. Abe, S. Hosaka, Y. Sakai, and H. Saito, A ^{13}C nuclear magnetic resonance study of covalently cross-linked gels. Effect of chemical composition, degree of cross-linking, and temperature to chain mobility, Macromolecules, 11:95 (1978).
4. F. Th. Hesslink, A Vrij, and J. Th. Overbeek, On the theory of the stabilization of dispersions by adsorbed macromolecules. II. Interaction between two flat particles, J. Phys. Chem., 75:2094 (1971).

5. D. J. Meier, Theory of polymeric dispersants. Statistics of constrained polymer chains, J. Phys. Chem., 71:1861 (1967).
6. N. G. Maroudas, Polymer exclusion, cell adhesion and membrane
   fusion, Nature, 254:695 (1975).
7. Y. Noishiki, An in vivo test for evaluation of blood compatibility by insertion of the polymer coated suture into the peripheral vein, Jpn. J. Artif. Organs, 11:794 (1982).
8. Y. Mori, S. Nagaoka, H. Takiuchi, T. Kikuchi, N. Noguchi,
   H. Tanzawa, and Y. Noishiki, A new antithrombogenic material
   with long polyethyleneoxide chains, Trans. Am. Soc. Artif.
   Intern. Organs, 28:459 (1982).
9. Y. Mori, S. Nagaoka, H. Takiuchi, R. Terada, S. Nishiumi,
   H. Tanzawa, A. Kuwano, and H. Miyama, The interaction between
   the hydrogels with polyethyleneoxide (PEO) chains and platelets,
   Artificial Organs, in press.
10. Dr. D. Regoli and V. Clark, Prevention by adenosine of the
    effect of adenosine diphosphate on the concentration of circu-
    lating platelets, Nature, 200:546 (1963).

USE OF METHYL CYANOACRYLATE (MCA) AS A SCLEROSING AGENT IN FEMALE
STERILIZATION: EFFECT OF INHIBITORS AND RADIOOPAQUE ADDITIVES ON
MCA POLYMERIZATION IN VITRO AND ON OVIDUCT OCCLUSION IN VIVO IN
RABBITS

James A. Nightingale, Allan S. Hoffman, Sheridan A.
Halbert, and Richard G. Buckles[+*]

Center for Bioengineering, Department of Chemical
Engineering and Department of Biological Structure
University of Washington, Seattle, Washington
*University of Basel, Basel, Switzerland

INTRODUCTION

In 1981 the world population was approximately 4.5 billion
and growing at the rate of about 80 million a year.[1] Many of the
developed countries have low or non-existent increases in population while the bulk of the expansion is occurring in developing
countries, primarily in a very few countries which include:
India, Indonesia, Brazil, Nigeria, Bangladesh, Pakistan, and
Mexico.[2] The rapid population growth that these countries are
experiencing results in a declining quality of life and a worsening
prognosis for those yet to come.[3]

Concerned people throughout the world seek to implement safe,
effective birth control techniques. This paper summarizes our
research to improve a method of birth control that induces non-
reversible female sterilization for consenting women (who usually
already have several healthy living children). A simple technique
combined with high effectiveness and suitable for non-surgical
theatres would be appropriate for developing nations. In the late
60's Neuwirth et al. began developing such a method using methyl
cyanoacrylate (MCA), applied as a sclerosing agent to cause
occlusion of the oviducts.[4]

The sclerosing agent is instilled using a special trans-
cervical delivery system that can be used by trained paramedics in
countries where surgical sterilization is not readily available.[5]
The MCA is instilled into the fundal region of the uterus, where
+ Deceased 10/21/81.

375

it mixes with available fluids. It is then forced into the oviducts by the action of an expanding balloon. The MCA polymerizes in seconds to minutes without spilling into the peritoneum. This poly(MCA) then degrades over a period of months, releasing cytotoxic degradation products that elicit a local inflammatory response, resulting in occlusion of the oviduct by fibrous (scar) tissue. Clinical studies produced unsatisfactory results.[6] These are summarized in Table 1.

When in contact with water, MCA is believed to polymerize by an anionic mechanism.[7] Coover attributed the incredible activity of the cyanoacrylates to the electromeric effects of both the nitrile and the alkoxycarbonyl groups. Figure 1 shows the proposed polymerization reaction. In the presence of a basic initiator these groups are rendered strongly electronegative. This stabilizes a charge distribution between the ethylene carbons where the unsubstituted carbon is left with an apparent positive charge and thus will attract anions. Almost any anion can initiate polymerization while acids tend to retard the reaction. We have developed several in vitro models that simulate the polymerization and mixing in vivo. Table 2 shows these in vitro models. The first three models are of increasing complexity and represent our desire to visualize the instillation process and understand the low efficacy observed in the clinic. We found that the polymerization process could be separated into two parts, interfacial and bulk, which were examined using the last two models.

The development of a radiopaque cyanoacrylate formulation was originated by clinicians who utilized iso-butyl cyanoacrylate to occlude arteries. The majority of contrast media approved for use in humans are soluble in and dissolved in water. The need for a material miscible with cyanoacrylate that will not initiate polymerization is evident. Cromwell and Kerber in 1979 formulated a radiopaque monomer by the addition of an iophendylate (Pantopaque, Lafayette Pharmacal) suitable for their work.[8]

Information from this work aided our development of a radiopaque additive. To examine this MCA/radiopaque formulation we undertook a limited number of in vivo tests using rabbits.

Table 1. Problems Encountered During Clinical Studies

1) Low success in bilateral occlusion of the oviducts (78%)

2) Potential peritoneal spillage (3/47 cases)

3) No method to visualize filling

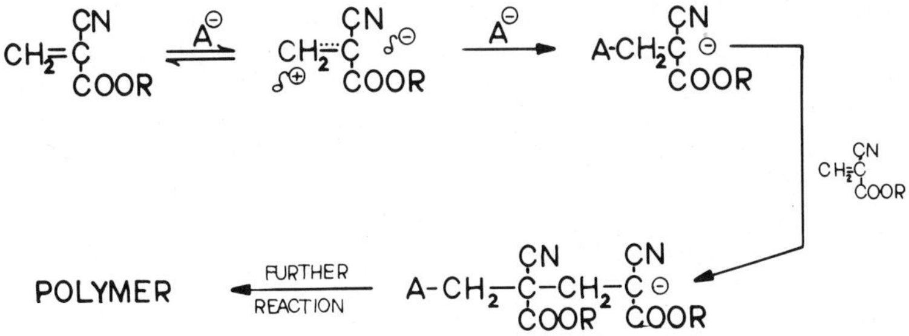

Fig. 1. Alkyl 2-Cyanoacrylate Polymerization Reaction.[7]

MATERIALS AND METHODS

Initially, we examined the instillation process where a monomer MCA jet contacts aqueous media. To visualize this mixing we constructed a small tank which could be filled with Hank's solution (a physiologic media) and heated to 37 C. A port on one end of the tank allowed rapid instillation of MCA into an immersed tube (inside the tank) by means of a pneumatic driven piston that was located outside the tank. We used this to examine the mixing and flow of MCA as it was instilled through an 18G flat tip needle into glass tubes (1 mm I.D.), cellophane tubes (Spectrum Medical Industry, 2.5 mm I.D.) and excised rabbit oviducts. The rabbit oviducts were removed quickly from a sacrificed animal and were surgically modified to simulate human physiology. They were then immersed into the bath and ligated to the injection port. Toagosei supplied the MCA used in this study; it contained hydroquinone (a free radical trap) and varying amounts of SO_2 (an acidic inhibitor).

To study the interfacial polymerization we used spreading MCA drops on physiologic media (Hank's). The pH of the media could be modified by the addition of either dilute HCl or aqueous $NaHCO_3$. A 5 μl drop of MCA, containing varying quantities of an acid inhibitor, was allowed to drop onto the surface of the media. The area of the spreading drop after polymerization was then observed.

Table 2. Models to Study MCA Instillation and Polymerization

NUMBER	MODEL
1	Glass tube immersed in Hank's solution
2	Cellophane tube immersed in Hank's solution
3	Freshly excised rabbit oviduct in Hank's solution
4	Spreading MCA drop on aqueous media
5	Falling ball in a cellophane tube of polymerizing MCA

377

Bulk polymerization was studied using a falling ball visco-
meter, which consisted of a cellophane tube (Spectrum) containing
monomer MCA and a 2.38 mm stainless steel ball. This was immersed
in Hank's solution pH 7.5. The tube was repeatedly inverted and
the time for the ball to fall between two marks measured as a
function of time.

The minimum quantity of radiopaque additive (Pantopaque)
for efficacy was determined using rabbit oviducts filled with
varying proportions of this iophendylate. These were exposed to
a standard x-ray dose using a phantom to simulate the body tissue
surrounding the oviduct. We also utilized selected tests from
those outlined above to investigate the effect of this additive
on polymerization.

A limited number of _in vivo_ tests were carried out using
proven doe rabbits. For each animal the radiopaque formulation
was instilled into an oviduct; the other oviduct then served as a
control and received 100% MCA. After a period of time these animals
were mated with males. Table 3 shows the components of the MCA
formulation studied.

DISCUSSION AND RESULTS

From the first experiment we realized that study of MCA
polymerization in a flow system would be difficult. The use of
glass tubes afforded a method to visualize this process using end
point examination and high speed photography. We found that two
distinct regions of poly(MCA) plug were formed after high speed
instillation. The first consisted of a "foamed" plug that was
porous and contained water, while the second was a solid "glassy"
polymer. Furthermore, as the time of injection was decreased

Table 3. Composition of MCA Formulation Studied
(MCA provided by Toagosei)

COMPONENT	FORMULA	NAME	TYPICAL CONCENTRATION
MONOMER	$CH_2 = C \begin{smallmatrix} CN \\ CO_2CH_3 \end{smallmatrix}$	METHYL CYANOACRYLATE (MCA)	(75 – 100%)
POLYMERIZATION INHIBITORS	SO_2	SULFUR DIOXIDE	100 ppm
	$CH_3O-\langle\rangle-OH$	METHOXY–HYDROQUINONE	100 ppm
RADIOPAQUE ADDITIVE	$I-\langle\rangle-CH-(CH_2)_8-CO_2C_2H_5$ with CH_3	IOPHENDYLATE ("PANTOPAQUE"-LAFAYETTE PHARMACAL)	(UP TO 25%)

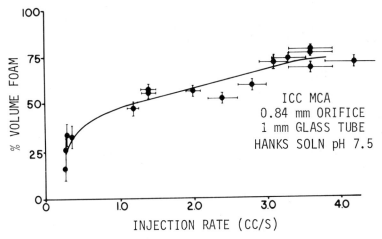

Fig. 2. Effect of Injection Date on Foamed poly(MCA) Formation and Monomer MCA Injected through an 18G Needle.

(increasing flow rate) there was an increase in the proportion of foamed poly(MCA) formed. Figure 2 shows the results from this study.

At injection rates higher than 1 cc/s the foam was seen at the distal end while at low injection rates the foam was dispersed in clumps randomly throughout the glassy poly(MCA) plug. The presence of the foam might be a result of entrainment of water by the initial MCA jetting out of the orifice into the tube. As the water is entrained a rapid interfacial polymerization occurs. Subsequent MCA entering the tube does not have access to water. The water has either been pushed ahead by the foam plug or the MCA has "skinned" over with a polymer coat which prevents entrainment.

These foamed plugs were also observed in freshly excised rabbit oviduct that had been injected with MCA under similar conditions. The clinical delivery system has an orifice I.D. of about 0.38 mm. Fluid flowing out of this small orifice would tend to form foamed-solid poly(MCA) as the jetting MCA mixes with fluids in the uterus.

The interfacial polymerization seemed to be the key to improving bilateral entry of the MCA into the oviducts. To study this we adopted a method used by Leonard to measure spreading of cyanoacrylate drops on fluids.[9] Figure 3 shows the effect of SO_2 concentration and media pH on spreading MCA drop area. It can be seen that large increases in area occur with increasing acid concentration in the drop or with decreasing media pH. The ability of the polymerizing MCA to spread on aqueous media may be closely

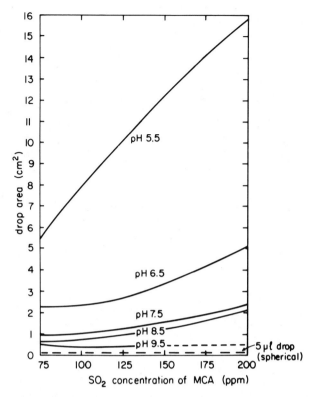

Figure 3.   Effect of SO$_2$ Concentration and Medium (Hank's) pH
on Spreading Drop Area.

related to the polymer film strength and polymerization rate at
the drop/medium interface. For highly inhibited MCA drops on
acidic media the poly(MCA) film is weak and allows the monomer to
spread. Conversely, the uninhibited polymerization reaction on pH
basic media rapidly forms a thick polymer film that resists the
spread of the monomer MCA.

Other organic acids have shown similar effects on spread drop
area. Figure 4 shows these results at medium pH = 7.5. Note that
the scale of this graph is considerably different than that shown
in the previous figure. Generally, as the molecular weight of the
carboxylic acid increases the spreading decreases.

If the monomer was inhibited to a very high degree we might
expect that it would be more likely to enter the oviduct. The
effect of high inhibitor levels might also increase the likelihood
of spillage into the peritoneum. We observed while using excised
rabbit oviducts that the peristaltic activity of the oviduct is
quite high. If the monomer does not polymerize promptly this

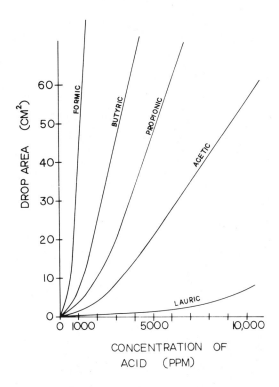

Figure 4.   Effect of Organic Acids on Spreading MCA (75 ppm SO_2)
Dropped on Hank's Solution pH 7.5, 25°C.

activity might force the monomer out the distal end of the oviduct
(and into the peritoneum, in vivo). To examine this bulk polymerization we developed a viscometric test. As the concentration of
SO_2 increased from 20 ppm to about 200 ppm there was an increase in
time to "set-up" (this was the time for the ball to completely
stop moving). Figure 5 shows the results from this series of
experiments. The plateau may be a result of diffusion of the
inhibitor out of the MCA into the aqueous medium. A drop in the
pH of the aqueous medium was also observed.

Note that the times for this experiment are quite long.
Results from an in vivo bulk polymerization test showed much shorter
times for polymerization in the oviduct after instillation. It was
found that for MCA containing 100 ppm SO_2 the time to set-up was
only one to two minutes. To determine these end points we used
digital palpation of the oviduct. This observation is probably

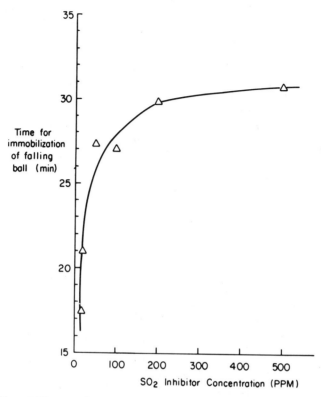

Figure 5.　Effect of SO$_2$ Contact on MCA Set-up Time (Cellophane
Tube Containing MCA is Immersed in Hank's Solution
pH 7.5, 25°C).

the result of the increased surface to volume ration and entrainment of aqueous fluids during instillation.

The minimum necessary quantity of the radiopaque additive for
radiographic purposes was determined to be 25% by volume Pantopaque.
This formulation resulted in reduced polymerization rates.　There
was an increase in observed drop area and in set-up times.　Figures
6 and 7 illustrate these results.

These results can be explained using Figure 8, where we see
that the Pantopaque can contain byproducts of manufacture that are
acids.　This acid could then act to inhibit polymerization.

Repeated matings of rabbits that had been instilled six
months previously with 100% MCA in one oviduct and 75% MCA/25%
Pantopaque in the other oviduct resulted in no pregnancies.　At
necropsy, the oviducts did, however, show a slightly higher degree
of hydrosalpinx in those that had been instilled with 100% MCA.

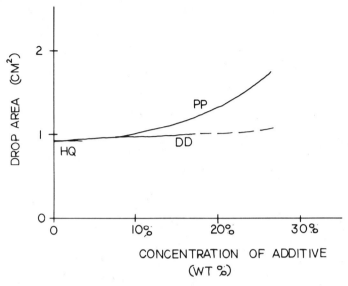

Figure 6. Effect of Additives on Spreading MCA (75 ppm SO_2)
Dropped on Hank's Solution.
DD: Dodecane
HQ: Hydroquinone
PP: Pantopaque

This condition is usually indicative of complete closure in
distal regions of the oviduct, allowing a build-up of fluids in
proximal portions, which swells the oviduct.

SUMMARY

Our results indicate that more attention should be focused on
injection pressures, environment of the uterus, and the geometry of
the injection system to minimize foaming which would waste monomer
and also possibly prevent entry of remaining MCA monomer through
the uterotubal junction into the oviduct. Reduction of the pH of
the uterus prior to instillation could inhibit polymerization in
the uterine cavity. Decreasing the injection rate would tend to
reduce foaming. Changes in formulation may improve the success
rates of bilateral occlusion without risking spillage into the
peritoneum. Several different organic acids can be used to
inhibit interfacial polymerization and may be used instead of SO_2.
The use of a radiopaque additive formulation shows promise. A
patent describing the use of radiopaque additives for this
application was assigned to the World Health Organization in 1982
as a result of this research.[10]

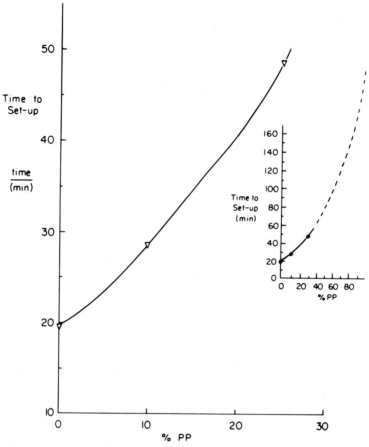

Figure 7. Effect of Radiopaque Additive on Cure Time for MCA
Filled Cellophane Tube Immersed in Hank's Solution
pH 7.5, 25°C.

        The authors would like to thank Elizabeth Zielie for
assistance in preparing this paper. We are particularly
indebted to the encouragement and assistance of the late
Dr. Richard G. Buckles. This investigation received financial
support from the Special Programme of Research, Development and
Training in Human Reproduction, World Health Organization,
Geneva, Switzerland.

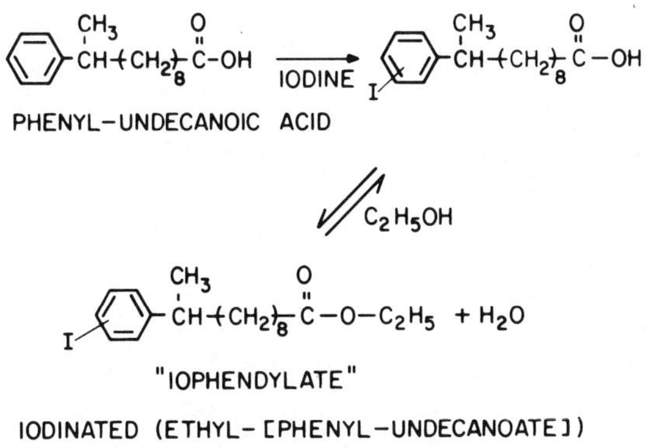

PHENYL–UNDECANOIC ACID

"IOPHENDYLATE"

IODINATED (ETHYL–[PHENYL–UNDECANOATE])

Figure 8. Synthesis and Impurities in "Iophendylate" (Pantopaque).

REFERENCES

1. "World Population: Toward the Next Century," Population
    Reference Bureau, Inc., Washington D.C., November 1981.
2. Sir M. Kendall, 1979, The World Fertility Survey: Current
    Status and Findings, Population Reports 7(4):M73.
3. R.M. Salas, 1981, "The State of World Population 1981,"
    United Nations Fund for Population Activities.
4. R.S. Neuwirth et al., 1968, An Outpatient Approach to Female
    Sterilization with Methylcyanoacrylate, Am. J. Obstet.
    Gynecol. 130:951.
5. R.M. Richart, R.S. Neuwirth and L.R. Bolduc, 1977, Single-
    Application Fertility Regulating Device: Description of a
    New Instrument, Am. J. Obstet. Gynecol. 127:86.
6. R.S. Neuwirth, R.M. Richart and L.R. Bolduc, 1977, Clinical
    Trials with the Single-Application Fertility Regulating
    Device, Am. J. Obstet. Gynecol. 129:348.
7. H.W. Coover, F.B. Joyner, N.H. Shearer and T.H. Wicker, 1959,
    Chemistry and Performance of Cyanoacrylate Adhesives, SPEJ
    15:413.
8. L.D. Cromwell and C.W. Kerber, 1979, Modification of Cyano-
    acrylate for Therapeutic Embolization: Preliminary Experience,
    AJR 132:799.
9. F. Leonard, J.A. Collins and H.J. Porter, 1966, Interfacial
    Polymerization of n-Alkyl-alpha Cyanoacrylate Homologues,
    J. Appl. Poly. Sci. 10:1617.
10. A.S. Hoffman, 1982, Method and Composition Containing MCA
    for Female Sterilization, U.S. Patent 4,359,454.